$\mathcal{S}$CIENCE WORKSHOP SERIES

Biology

SURVEY OF LIVING THINGS

with an Introduction to Scientific Methods

Seymour Rosen

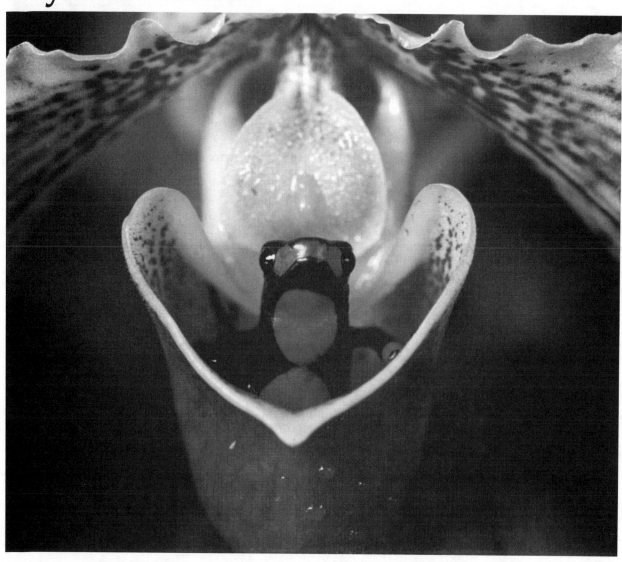

GLOBE FEARON

Pearson Learning Group

THE AUTHOR

Seymour Rosen received his B.A. and M.S. degrees from Brooklyn College. He taught science in the New York City School System for twenty-seven years. Mr. Rosen was also a contributing participant in a teacher-training program for the development of science curriculum for the New York City Board of Education.

Cover Designer: Joan Jacobus
Cover Photograph: Art Wolfe/Tony Stone Images
Cover Photo Researcher: Joan Jacobus
Photo Researchers: Rhoda Sidney, Jenifer Hixson

About the cover illustration: Orchids are one of the oldest known flowery plants. At one time, they were believed to be parasitic because they grew on trees. Later people discovered that this was not true. Orchids lived on trees, but did not feed off them. Today, there are around 35,000 species in existence. The Poison Arrow Frog, also known as the Poison Dart Frog, produces more than 200 heart and nerve toxins so they won't be eaten by predators.

Photo Credits:
p. 33, Fig. A: Helena Frost
p. 33, Fig. B: Al Ruland
p. 34, Fig. E: Czechoslovakia Tourist Office
p. 34, Fig. F: Salt River Project
p. 35, Fig. G: Province of Quebec Film Bureau
p. 53, Fig. C: Harry E. Mopsikoff
p. 53, Fig. D: U.S. Department of Agriculture
p. 53, Fig. E: U.S. Department of Agriculture
p. 54, Fig. F: William A. Frost
p. 55, Fig. G: William A. Frost
p. 55, Fig. H: William A. Frost
p. 57, Fig. I: William A. Frost
p. 57, Fig. J: William A. Frost
p. 58, Fig. K: Photo Researchers
p. 59, Fig. L: William A. Frost
p. 59, Fig. M: William A. Frost
p. 59, Fig. N: William A. Frost

p. 60, Fig. O: William A. Frost
p. 70: Peter Vandermark/Stock, Boston
p. 93, Fig. H: Alvin E. Staffan/Photo Researchers
p. 113, Fig. I: Dan Guravich/Photo Researchers
p. 142, Fig. D: Barry L. Runk/Grant Heilman
p. 143, Fig. J: Blair Seitz/Photo Researchers
p. 150, Fig. E: Richard Carlton/Photo Researchers
p. 158: Laurence Pringle/Photo Researchers, Inc.
p. 209, Fig. A: Runk/Schoenberger, Grant Heilman
p. 209, Fig. B: Grant Heilman
p. 209, Fig. D: UPI/Bettmann Newsphotos
p. 212, Fig. H: Michael Heron
p. 218: Rhoda Sidney

ISBN: 0-130-23378-1

Printed in the United States of America

11 12 13 14 15 V039 10

Globe
Fearon

Pearson Learning Group

1-800-321-3106
www.pearsonlearning.com

CONTENTS

Introduction to Biology

Have you ever wondered what you were made up of? The ancient Greeks believed everything was made up of four substances: earth, air, fire, and water. However, with the invention of the microscope, a new world was discovered. People found out that cells made up living things.

Using the microscope, scientists could observe microscopic living things that they had never seen before. They even found that hundreds of organisms could be found in one drop of water!

In this book, you will study biology, the study of living things. You will learn about the simplest, single-celled organisms, such as monerans, protists, and fungi. You also will learn about more complex, multicelled organisms, such as plants and humans.

However, most importantly, when you finish this book you will have a basic understanding of the world around you.

Lesson

How do scientists measure things?

TRIPLE BALANCE BEAM

KEY TERMS

mass: amount of matter in an object

weight: measure of the pull of gravity on an object

length: distance between two points

area: measure of the size of a surface

volume: measure of the amount of space an object takes up

temperature: measure of how hot or cold something is

LESSON 1 | How do scientists measure things?

How much do you weigh? What is your height? How many tiles will cover your kitchen floor? How much milk should be added to a cake mix? What is the temperature outside? All of these questions are answered by measurements.

Measuring is an important part of daily life. People use measurements all the time—for shopping, cooking, construction, and deciding how warmly to dress. Measuring also is an important part of science.

A measurement has two parts: a <u>number</u> and a <u>unit</u>. A unit is a standard amount used to measure something.

EXAMPLES
$$\underset{\text{number} \qquad \text{standard}}{100 \text{ grams}} \qquad \underset{\text{number} \qquad \text{standard}}{25 \text{ liters}}$$
$$\text{unit} \qquad\qquad\qquad \text{unit}$$

There are many kinds of measurements. The most common are:

MASS and weight are related, but they are not the same. **Mass** is a measure of the amount of matter in an object. **Weight** is a measure of the pull of gravity on an object. The basic unit of mass in the metric system is the kilogram (kg). Mass is measured with a balance.

LENGTH is the distance from one point to another as measured by a ruler. The basic metric unit of length is the <u>meter</u> (m). You can use a meter stick or metric ruler to measure length.

AREA is a measure of <u>surface</u> room—how big something is in two directions. You can find the area of a rectangle by multiplying its length by its width. Area is measured in square units, such as square meters (m^2).

VOLUME is the measure of the amount of <u>space</u> an object takes up—how big an object is in all three directions. The <u>liter</u> (L) is the basic unit of volume in the metric system. A <u>measuring cup</u> or a <u>graduated cylinder</u> is used to measure the volume of liquids.

The volume of a solid can be measured in cubic centimeters (cm^3). You can find the volume of a cube or rectangle by multiplying its length by its width by its height. 1000 cubic centimeters equals 1 liter.

TEMPERATURE is the measure of how hot or cold an object is. Temperature is measured with a thermometer in <u>degrees Celsius °C</u>, or <u>degrees Fahrenheit °F</u>. The Celsius scale usually is used in science.

UNDERSTANDING METRICS

In the United States, people usually use <u>English</u> units of measurement such as <u>ounces</u>, <u>pounds</u>, <u>inches</u> and <u>feet</u>. Most other countries use metric units. Metric units include the <u>gram</u>, <u>kilogram</u>, <u>meter</u>, and <u>centimeter</u>. Scientists also use the metric system. In science, you will use mostly metric units.

The metric system is based upon units of <u>ten</u>. Each unit is ten times smaller or larger than the next unit. This means that you can convert a measurement from one unit to another by multiplying or dividing by ten. Prefixes describe a unit's value. The prefixes and their meanings are listed below.

PREFIX	MEANING	
kilo- [KILL-uh] ————	one thousand (1,000)	} each larger by a multiple of <u>ten</u>
hecto- [HEC-tuh]————	one hundred (100)	
deca- [DEC-uh] ————	ten (10)	
deci- [DESS-ih] ————	one tenth (1/10)	} each smaller by a multiple of 1/10
centi- [SEN-tih] ————	one hundredth (1/100)	
milli- [MILL-ih] ————	one thousandth (1/1,000)	

Use the chart above to answer the following questions.

1. How many grams make up a kilogram? _____1000_____
   10, 100, 1,000

2. How much of a meter is a centimeter? _____1/100_____
   1/10, 1/100, 1/1,000

3. How many times larger is a hectometer compared to a decameter? _____10_____
   10, 100, 1,000

4. How many times smaller is a millimeter compared to a decimeter? _____100_____
   10, 100, 1,000

5. Which prefix stands for a <u>greater</u> value?

   a) deca- or kilo-? _____   **d)** hecto- or kilo-? _____

   b) kilo- or milli-? _____   **e)** centi- or deci-? _____

   c) centi- or milli-? _____   **f)** deca- or deci-? _____

MEASURING MASS

1. In the metric system, the unit of mass is the _____ .

meter, kilogram, pound

2. Mass and weight _____ the same.

are, are not

3. _____ is a measure of the amount of matter in an object.

mass, weight

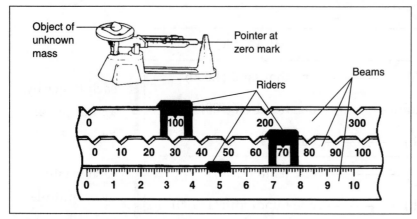

Figure A

4. What instrument is used to measure mass? _____

5. What is the mass of the object shown? _____

TRUE OR FALSE

In the space provided, write "true" if the sentence is true. Write "false" if the sentence is false.

_____   1. Weight is a measure of the pull of gravity on an object.

_____   2. Scientists use English units of measurement.

_____   3. The prefix centi- stands for one hundredth (1/100).

_____   4. A graduate is used to measure mass.

_____   5. The basic unit of length in the metric system is the meter.

_____   6. Volume is a measure of the amount of matter in an object.

_____   7. One kilogram is less than one gram.

_____   8. A measurement has two parts.

_____   9. A unit is an amount used to measure something.

_____   10. Most countries use the metric system.

MEASURING LENGTH

Length is measured with a metric ruler. Part of a combined metric and inch ruler is shown in Figure B. On the metric side of the ruler the distance between numbered lines is equal to one centimeter. Each centimeter is divided into 10 equal parts. Each one of these parts is equal to one millimeter.

The figure below shows a combined metric and inch ruler.

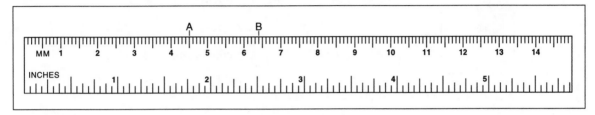

Figure B

1. What value does the prefix <u>milli-</u> stand for? _____

2. What value does the prefix <u>centi-</u> stand for? _____

3. Which is <u>larger</u>, a meter or a millimeter? _____

4. How many millimeters make 1 centimeter? _____

5. The length at A may be written as 45 mm. It may also be written as _____.
   45 cm, 4.5 cm, 4.5 mm

6. The length at B may be written as _____ mm or _____ cm.

Measure each of the following lengths. Write the lengths on the right in centimeters and millimeters.

7. ————————————————————— 7. _____ cm _____ mm

8. ————————————————————— 8. _____ cm _____ mm

9. ——————— 9. _____ cm _____ mm

10. ——————————————————————— 10. _____ cm _____ mm

To the right of each length listed, <u>draw</u> a line of that length.

a) 92 mm

b) 9.2 cm

c) 43 mm

d) 3.5 cm

5

MEASURING AREA

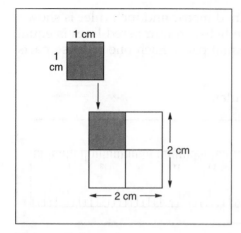

Figure C

The larger square in Figure C has an area of 4 square centimeters (4 cm²).

Area = length × width

= 2 cm × 2 cm

Area = 4 square centimeters (4 cm²).

Figure the area of each of the following rectangles: (Measure Figures G and H yourself.)

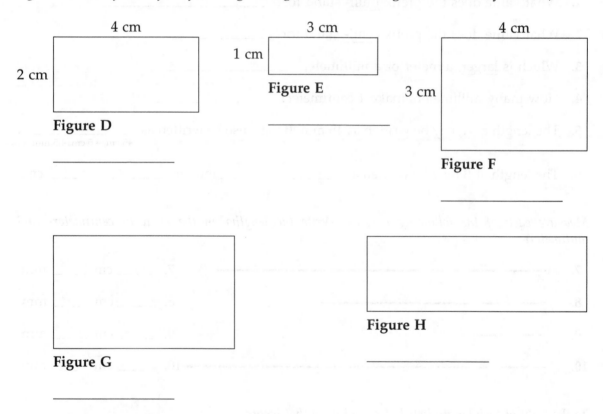

4 cm

2 cm

Figure D

_____

3 cm

1 cm

Figure E

_____

4 cm

3 cm

Figure F

_____

Figure G

_____

Figure H

_____

CALCULATING AREA

Find the volume of the following rectangles:

1. 5 meters × 5 meters _____

2. 2.5 cm × 5 cm _____

3. 10 millimeters × 10 millimeters _____

The volume of liquids is measured in a <u>graduated cylinder</u>. A graduated cylinder is a glass tube that is marked with divisions to show the amount of liquid in it. To measure liquid volume, you hold the graduated cylinder at your <u>eye level</u>. The surface of the liquid will have a "belly-down" curve. You should read the mark that lines up with the <u>bottom</u> of the curve.

What is the liquid volume in this graduated cylinder?

_____

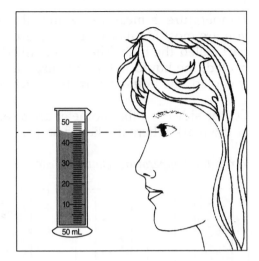

Figure I

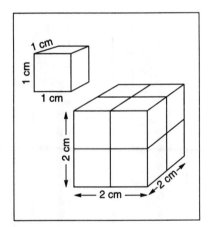

Figure J

What is the volume of a cube that is 2 cm × 2 cm × 2 cm?

Volume = l × w × h

= 2 cm × 2 cm × 2 cm

Volume = 8 cubic centimeters (8 cm³)

Find the volume of each of the following boxes:

Volume

1. 2 cm × 5 cm × 1 cm   _____

2. 8 m × 2 m × 2 m   _____

3. 1 mm × 1 mm × 10 mm   _____

4. 4 cm × 2 cm × 3 cm   _____

5. 5 m × 3 m × 6 m   _____

READING A CELSIUS THERMOMETER

Temperature is measured with a thermometer. Many thermometers, including the ones you are most familiar with, are made of glass tubes. At the bottom of the tube is a wider part called the bulb. The bulb is filled with a liquid, such as mercury. When the bulb is heated, the liquid in the bulb expands, or gets larger. It rises in the tube. When the bulb is cooled, the liquid contracts, or gets smaller. It falls in the tube.

On the sides of a thermometer are a series of marks. You read the temperature by looking at the mark where the liquid stops.

Write the temperature shown on each celsius thermometer in the space provided.

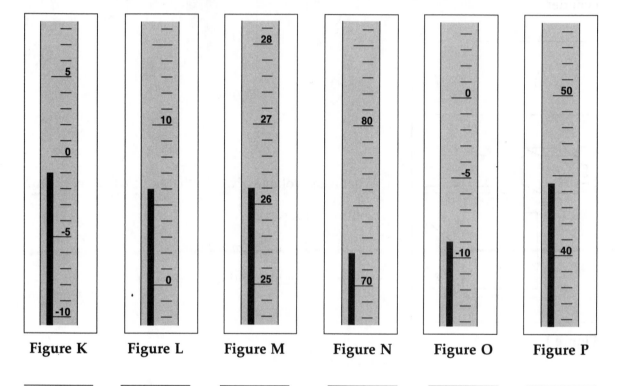

Figure K	Figure L	Figure M	Figure N	Figure O	Figure P
_____	_____	_____	_____	_____	_____

REACHING OUT

One cubic centimeter is equal to one millimeter (mL). How many <u>liters</u> of water can a

1,800 cm³ pan hold? _____

8

What is scientific method?

KEY TERMS

scientific method: problem solving guide

hypothesis: suggested solution to a problem based upon known information

data: record of observations

LESSON 2 | What is scientific method?

You may not realize it, but you do problem-solving every day. You do not always think about how to solve a particular problem. You solve problems in sort of a "natural" way, a way that seems to "make sense." And it usually does.

For example, suppose you put your key into your house door and try to turn it, but it does not budge. You wonder what's wrong. You examine the key to make certain that it is the correct one. Then you try again. The key still does not turn. What next? You might "jiggle" the key. Or, you might pull back on the doorknob as you try to turn the key. One of these approaches <u>might</u> work. If not, you try other methods until the problem is solved.

Without knowing it, you solve problems very much like a scientist does. You use **scientific method.** Scientific method is a guide used to solve problems. It involves asking questions, making observations, and trying things out in an orderly way. Scientists use certain steps to solve problems. The steps of scientific method are:

- **IDENTIFY THE PROBLEM** State it clearly—usually as a question.

- **GATHER INFORMATION** Research; ask questions. Discover what is already known about the problem.

- **STATE A HYPOTHESIS** A **hypothesis** [hy-PAHTH-uh-sis] is a suggested solution as to why something happens.

- **TEST THE HYPOTHESIS** Experiment and examine the situation to check the hypothesis.

- **MAKE CAREFUL OBSERVATIONS** Note everything your senses can gather. Record the **data** [DAYT-uh]. Keep careful records.

- **ORGANIZE AND ANALYZE THE DATA** Put the data in order. Scientists often use charts and tables to organize data. Figure out the <u>meaning</u> of the data.

- **STATE A CONCLUSION** Explain the data. State whether or not it supports the hypothesis.

Different problems require different approaches. Not every step in scientific method needs to be used. And the steps can be used in any order.

CHOOSE THE RIGHT CAPTION

On this page and the next, there are eight figures and eight captions. Each caption matches one of the figures. Choose the caption that best describes each figure. Write the correct caption on the line provided.

Choose from these captions:

Identify the problem   Make careful observations

Gather information   Record the data

State a hypothesis   Analyze the data

Test the hypothesis   State a conclusion

Figure A

1. _____

Figure B

2. _____

Figure C

3. _____

Figure D

4. _____

Figure E

5. _____

Figure F

6. _____

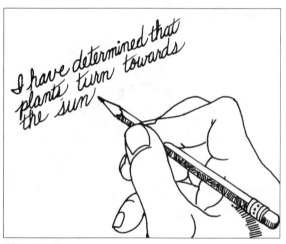

Figure G

7. _____

Figure H

8. _____

MATCHING

Match each term in Column A with its description in Column B. Write the correct letter in the space provided.

	Column A		Column B
_____	**1.** analyze	**a)**	explains the data
_____	**2.** scientific method	**b)**	suggested solution
_____	**3.** conclusion	**c)**	test the hypothesis
_____	**4.** hypothesis	**d)**	guide for solving problems
_____	**5.** experiment	**e)**	figure out the meaning

Two separate stories are shown in the figures below. However, the figures in each are not in the proper order (sequence). In the table under each set of figures, list the figures in their proper order. Also, explain what is happening in each figure. Finally, write a hypothesis (in question form) and a conclusion.

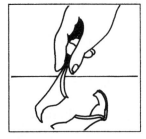
Figure I

Figure J

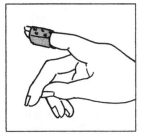
Figure K

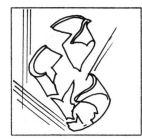
Figure L

Step	Figure	Explanation
1.		
2.		
3.		
4.		

Hypothesis: _____

Conclusion: _____

Figure M

Figure N

Figure O

Figure P

Step	Figure	Explanation
1.		
2.		
3.		
4.		

Hypothesis: _____

Conclusion: _____

13

FILL IN THE BLANK

Complete each statement using a term or terms from the list below. Write your answers in the spaces provided.

supports	observe	different
problems	question	already known
data	scientific method	senses
steps		

1. To test a hypothesis, scientists may _____ natural events.

2. When scientists research, they may find out what is _____ about a problem.

3. Your _____ gather information.

4. A conclusion states whether or not data _____ a hypothesis.

5. A problem is usually stated as a _____ .

6. Scientists use certain _____ to solve problems.

7. You solve _____ much like scientists do.

8. Different problems can be solved in _____ ways.

9. A guide used to solve problems is called _____ .

10. Scientists use charts to put _____ in order.

REACHING OUT

Jennifer has never eaten asparagus. She is afraid that it might make her sick. At dinner, she eats some. She likes the taste, but soon she suffers from nausea. Jennifer concludes that asparagus makes her sick.

1. Why might Jennifer's conclusion be <u>incorrect</u>?_____

_____

_____

2. What might be done to further test her conclusion? _____

_____

_____

How are experiments done safely?

KEY TERM

safety alert symbols: signs that warn of hazards or dangers

LESSON 3 | How are experiments done safely?

"Hands-on" experiences are part of many school activities. Science, especially, is suited to "learning by doing." You investigate; you make things happen; you learn from what you do.

Science investigations can be exciting. However, they can also be dangerous. Science laboratories have equipment and materials that can be dangerous if not handled properly. For this reason, it is important for you to always follow proper safety guidelines. Safety rules are for your own protection—as well as the protection of everyone around you.

The safety rules that you should follow are listed on page 17. Read over these safety rules carefully. Notice the **safety alert symbols** that accompany the safety rules. In this book, safety alert symbols are included at the beginning of some activities to make you aware of safety precautions. Always note any safety symbols and caution statements in an activity.

To avoid accidents in the science laboratory, always follow your teacher's directions. You should not perform activities without directions from your teacher. You also should never work in the science laboratory alone.

One hazard has no symbol even though it probably causes more accidents than any others. That hazard is "horsing around." Horsing around in the laboratory can lead to serious injury—or even death. So THINK before doing anything foolish.

CLOTHING PROTECTION • A lab coat protects clothing from stains. • Always confine loose clothing.

 EYE SAFETY • Always wear safety goggles. • If anything gets in your eyes, flush them with plenty of water. • Be sure you know how to use the emergency wash system in the laboratory.

 FIRE SAFETY • Never get closer to an open flame than is necessary. • Never reach across an open flame. • Confine loose clothing. • Tie back loose hair. • Know the location of the fire-extinguisher and fire blanket. • Turn off gas valves when not in use. • Use proper procedures when lighting any burner.

 POISON • Never touch, taste, or smell any unknown substance. Wait for your teacher's instruction.

 CAUSTIC SUBSTANCES • Some chemicals can irritate and burn the skin. If a chemical spills on your skin, flush it with plenty of water. Notify your teacher without delay.

 HEATING SAFETY • Handle hot objects with tongs or insulated gloves. • Put hot objects on a special lab surface or on a heat-resistant pad; never directly on a desk or table top.

 SHARP OBJECTS • Handle sharp objects carefully. • Never point a sharp object at yourself—or anyone else. • Cut in the direction away from your body.

 TOXIC VAPORS • Some vapors (gases) can injure the skin, eyes, and lungs. Never inhale vapors directly. • Use your hand to "wave" a small amount of vapor towards your nose.

 GLASSWARE SAFETY • Never use broken or chipped glassware. • Never pick up broken glass with your bare hands.

 CLEAN UP • Wash your hands thoroughly after any laboratory activity.

 ELECTRICAL SAFETY • Never use an electrical appliance near water or on a wet surface. • Do not use wires if the wire covering seems worn. • Never handle electrical equipment with wet hands.

 DISPOSAL • Discard all materials properly according to your teacher's directions.

PUTTING SAFETY RULES TO USE

Answer the following questions in complete sentences.

1. Jean has long hair. What should she do before working near an open flame? _____

   _____

2. A glass tube has broken. How should you pick up the pieces? _____

   _____

3. Why should you always wear safety goggles during <u>every</u> lab activity? _____

   _____

4. What else should you wear? Why? _____

   _____

5. A chemical spills on your hand. You are pretty sure that it is harmless. But you are

   not certain. What should you do? _____

IDENTIFYING SAFETY ALERT SYMBOLS

Six safety alert symbols are shown below. Match them with their meanings. Write the correct <u>letter</u> next to each description.

a.   **b.**   **c.**   **d.**   **e.**   **f.**

_____ 1. electrical safety   _____ 4. clothing protection

_____ 2. fire safety   _____ 5. sharp objects

_____ 3. heating safety   _____ 6. glassware safety

REACHING OUT

In the box at the right, design a NO HORSING AROUND symbol. Either draw it or describe it, or both. Perhaps you can think up more than one.

Lesson 4

How do we know when something is alive?

KEY TERMS

organisms: living things

response: a reaction to a change in the environment

adaptation: characteristic of an organism that helps the organism survive

stimulus: a change in the environment

LESSON 4 | How do we know when something is alive?

The world around you is made up of many different things. Some things, such as plants and animals, are living. Other things, such as cars and rocks, are nonliving.

Living things are called **organisms** [AWR-guh-nizms]. You know that plants and animals are living things. How do you know this? All living things have the following characteristics:

CELLS Living things are made up of one or more cells. Cells are often called the "building blocks of life."

ENERGY All organisms use energy. You get energy from the food you eat.

RESPONSE Living things react or **respond** to changes in the environment. For example, a sudden loud noise can cause you to jump. When the sun rises, the petals of a flower open.

ADAPTATION Organisms are **adapted** [uh-DAPT-ed], or suited, to their surroundings. For example, polar bears live in sub-zero temperatures. You and most other organisms cannot. Polar bears are adapted to the bitter cold. They have a thick layer of fat and dense fur.

REPRODUCTION [ree-proh-DUK-shun] Living things make more of their own kind. Organisms can reproduce only their own kind. Mice give birth only to mice. Elm trees produce new elm trees. Roses reproduce only roses.

GROWTH AND DEVELOPMENT Living things change or develop, during their lifetimes. One way organisms change is by growing bigger. You have grown since you were born. When living things grow they grow from the inside. Only living things can grow by themselves.

Anything that causes change, or activity in an organism is a **stimulus** (plural: stimuli). Stimuli come from both inside and outside of an organism. The change or activity caused by a stimulus is a **response**. A stimulus is like a message. A response is like an answer to the message.

Each picture shows a stimulus and response. What are the stimulus and response shown in each picture? Write your answers on the lines under each picture.

Figure A

Stimulus _____

Response _____

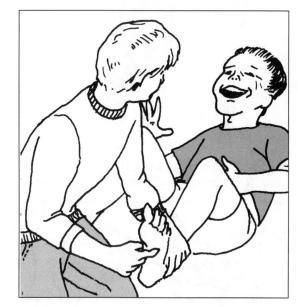

Figure B

Stimulus _____

Response _____

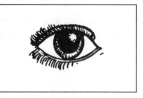

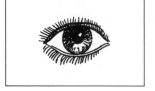

Figure C

These three pictures show how the pupils of the eyes react to light. Light is a stimulus. The top picture shows how your pupil might look in a dark room. The bottom picture shows how your pupil might look in bright light. The middle picture shows the eye in normal light.

1. In bright light, what happens to the size of your pupils?

    _____

2. In dim light, what happens to the pupil size? _____

    _____

3. How do you think this reaction by the pupil helps a

    person? _____

MORE ABOUT ADAPTATION

Any characteristic that helps an organism live in its environment is called an adaptation. Adaptations may take many forms. Some examples are shown in the pictures below.

Figure D *A woodpecker has a strong, pointy beak. It is adapted to dig insects out of trees.*

Figure E *Birds have feathers and lightweight bones. They are well adapted for flight.*

Figure F *The streamlined shape of a fish allows it to move quickly through the water.*

Figure G *A cactus has a thick, leathery stem that stores water. It is adapted to a dry, desert environment.*

MATCHING

Match each organism listed in Column A with the environment to which it is most suited to in Column B. Write the correct letter in the space provided.

	Column A	Column B
_____	**1.** Polar Bear	**a)** Forest
_____	**2.** Deer	**b)** Desert
_____	**3.** Goldfish	**c)** Soil
_____	**4.** Cactus	**d)** Water
_____	**5.** Earthworm	**e)** Cold, snowy climate

22

An important human adaptation is the thumb. How important is the thumb? More important than you may realize . . .

Figure H

Try this:

Tape your right thumb (if you are right-handed) to your next finger. Then try to do these tasks.

1. Fasten (or unfasten) a button.

2. Pick up a book.

3. Turn a doorknob.

4. Hold an object as you would a hammer.

5. Turn a screwdriver.

How important is the thumb? <u>You</u> answer the question!

Do you think civilization would be as advanced if people did not have thumbs? _____

_____

Explain your answer _____

_____

_____

FILL IN THE BLANK

Complete each statement using a term or terms from the list below. Write your answers in the spaces provided. Some words may be used more than once.

energy   adapted   respond
cells   reproduce organisms
stimulus

1. Living things are made up of one or more _____ .

2. All living things _____ their own kind.

3. Living things are called _____ .

4. Anything that causes change or activity in an organism is called a

   _____ .

5. You get _____ from food.

6. Living things _____ to changes in the environment.

7. Roses _____ only roses.

8. _____ are called the "building blocks of life."

9. Polar bears are _____ to the bitter cold.

TRUE OR FALSE

In the space provided, write "true" if the sentence is true. Write "false" if the sentence is false.

_____   **1.** Stimuli can come only from outside an organism.

_____   **2.** In dim light, your pupils get smaller.

_____   **3.** Only living things can grow by themselves.

_____   **4.** A response is like an answer to a message.

_____   **5.** Plants reproduce.

_____   **6.** All living things are made up of more than one cell.

_____   **7.** The thick stem of a cactus helps it survive in the desert.

_____   **8.** All living things use energy.

_____   **9.** Only animals are made up of cells.

What are the life processes?

KEY TERMS

ingestion: taking in of food

digestion: breaking down of food into usable forms

respiration: process by which organisms release energy from food

excretion: getting rid of waste products

circulation: movement of products throughout an organism

LESSON 5 | What are the life processes?

Organisms must carry on certain processes to stay alive. These processes are called <u>life processes</u>. The life processes also are characteristics of all living things.

A list of the life processes follows:

INGESTION An animal cannot make its own food. Animals must take in food and water from the outside. This life process is called ingestion [in-JES-chun]. Ingestion is the taking in of food.

Plants are different. Plants do not take in their food from the outside. Plants can make their own food. However, plants take in some important materials, such as water, from the soil.

DIGESTION Digestion [dy-JES-chun] is the breaking down of food into a form that can be used. Both plants and animals must digest their food. Plants and animals make special chemicals that change food into usable form.

In many animals, digestion takes place mainly in the stomach and small intestine. In most plants, digestion takes place where the food is made, in the leaves.

RESPIRATION Respiration [res-puh-RAY-shun] is the release of energy by combining oxygen with digested foods. During respiration, oxygen is used to break food apart. This process produces energy. Water and carbon dioxide also are produced. They are the waste products of respiration.

EXCRETION Excretion [ik-SKREE-shun] is the getting rid of waste products. Your body makes many waste products. These waste products must be removed from the body.

CIRCULATION Circulation [sur-kyuh-LAY-shun] is the movement of useful products to all parts of an organism and the carrying away of waste products. Circulation also is called transport.

WHICH LIFE FUNCTION?

Which life function do you see in each picture? Write the life function on the line below the picture.

Figure A_____

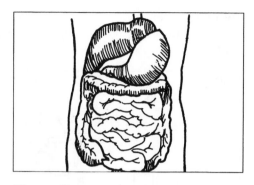

Figure B _____

Figure C_____

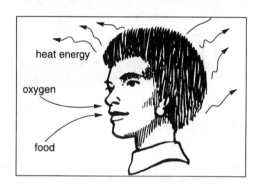

Figure D_____

Figure E _____

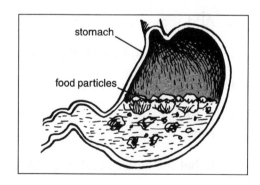

Figure F _____

Figure G_____

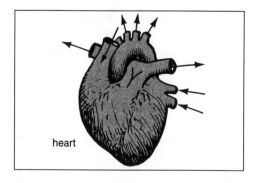

Figure H_____

What You Need (Materials)

two test tubes   digestive chemical
test tube rack   egg white from a hard-boiled egg
water

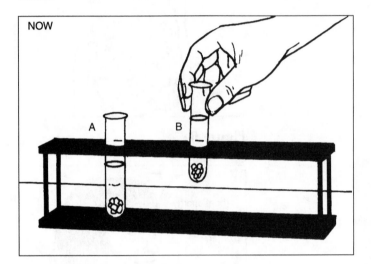

Figure I

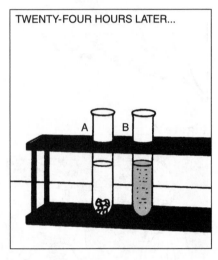

Figure J

How To Do The Experiment (Procedure)

1. Place a piece of egg white from a hard-boiled egg in a test tube.

2. Add some water to this test tube and label the test tube "A."

3. Place the test tube in the test tube rack.

4. Place another piece of egg white in a second test tube.

5. Add some water to this test tube. Then add some digestive chemical to this same test tube.

6. Label the second test tube "B" and put it into the test tube rack.

7. Let both test tubes stand overnight.

What You Learned (Observations)

1. The egg white in test tube "A" _____ change.
    did, did not

2. The egg white in test tube "B" _____ change.
    did, did not

3. Digestion is taking place in test tube _____ .
    A, B

4. The egg white is changing to a _____ .
    solid, liquid

28

5. Water alone _____ digest food.
   does, does not

6. What is in test tube "B" that is not in test tube "A"? _____

Something To Think About (Conclusions)

Which life process is happening in test tube "B"? _____

TRUE OR FALSE

In the space provided, write "true" if the sentence is true. Write "false" if the sentence is false.

_____ 1. Some animals can live without food.

_____ 2. Animals must ingest food.

_____ 3. Plants can make their own food.

_____ 4. Both plants and animals must digest their food.

_____ 5. A plant takes in water through its leaves.

_____ 6. Ingestion is the same as digestion.

_____ 7. When food is digested, it changes to a usable form.

_____ 8. After digestion, a steak is still a steak.

_____ 9. Every living thing has a stomach.

_____ 10. In plants, most digestion takes place in the leaves.

MATCHING

Match each term in Column A with its description in Column B. Write the correct letter in the space provided.

Column A

_____ 1. excretion

_____ 2. digestion

_____ 3. circulation

_____ 4. respiration

_____ 5. ingestion

Column B

a) produces energy

b) movement of products throughout an organism

c) taking in of food

d) getting rid of wastes

e) breakdown of food

MULTIPLE CHOICE

In the space provided, write the letter of the word that best completes each statement.

_____ 1. Oxygen is used to break food apart during
  a. ingestion.
  b. digestion.
  c. respiration.
  d. circulation.

_____ 2. In many animals, digestion takes place mainly in the stomach and
  a. small intestine.
  b. large intestine.
  c. lungs.
  d. kidneys.

_____ 3. Getting rid of waste products is called
  a. ingestion.
  b. digestion.
  c. circulation.
  d. excretion.

_____ 4. Transport is another name for
  a. digestion.
  b. circulation.
  c. ingestion.
  d. excretion.

_____ 5. Two waste products of respiration are
  a. water and oxygen.
  b. oxygen and carbon dioxide.
  c. water and carbon dioxide.
  c. food and water.

REACHING OUT

Excretion removes waste products from the body. What do you think would happen if waste products built up in the body? _____

_____

_____

What do living things need to stay alive?

LESSON 6 | What do living things need to stay alive?

Imagine that you are chosen to go to the moon. What would you take along? Food and water? Yes, of course. But is that all?

There is no oxygen in space. Where the sun shines, it is very, very hot. Where there is no sunlight, it is unbearably cold. On the moon, you would have to wear a spacesuit to protect you from the temperature extremes. You also would need an oxygen supply.

What must you take along on a trip to the moon? In short—your own environment.

As you know, living things are in constant contact with their environment. On Earth, the environment provides organisms with all the things they need to stay alive. For example, all living things need food, oxygen, water, and a proper temperature. Plants also need carbon dioxide. How is each of these things important?

FOOD   Food provides organisms with the energy for life. It also provides organisms with the materials needed for growth and repair.

Animals take in their food from the outside by eating plants or other animals. Plants make their own food.

AIR   Air is a mixture of gases. Oxygen is one of the gases in air. Oxygen is used by most living things for respiration. During respiration, oxygen breaks food apart. This produces energy.

Land organisms get most of their oxygen from the air. Organisms that live in the water take in oxygen that is dissolved in the water.

Carbon dioxide is another gas in air. Plants need carbon dioxide to make their own food.

WATER   All living things are made up mostly of water. In fact, between 65 percent and 95 percent of an organism's body may be made up of water. The materials needed for the life processes are dissolved in this water.

PROPER TEMPERATURE   Organisms live in many different climates. Some organisms live in hot climates. Others live in cold climates. The temperature of the environment is important to all living things.

Look at the pictures. Then answer the questions.

1. Where do animals get their food?

   _____

   _____

Figure A

2. Do all animals eat the same kind

   of food? _____

3. Why do organisms need food?

   _____

Figure B

4. Do plants take in food from the

   outside? _____

5. Where do plants get their food?

   _____

   _____

6. What gas do plants need to make

   their own food? _____

   _____

7. What part of a plant takes in water?

   _____

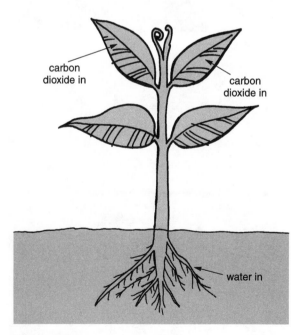

carbon dioxide in

carbon dioxide in

water in

Figure C

33

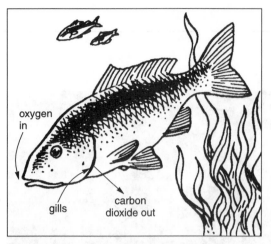

Figure D

Figure E

Figure F

8. The oxygen that fish breathe in is dissolved in the _____ .

9. Fish take in oxygen through _____ .
   gills, lungs

10. Living things need oxygen for the process of _____ .

11. During respiration, fish give off _____ as a waste product.
    oxygen, carbon dioxide

12. Why must all living things have water? _____
_____

13. How do most animals ingest water?
_____

14. Why do many desert plants have roots that reach deep into the ground?
_____
_____

WARMBLOODED AND COLDBLOODED ANIMALS

Some animals are coldblooded and some are warmblooded. The body temperature of a coldblooded animal changes with the temperature of the animal's surroundings. Fish are coldblooded animals. Frogs and toads are coldblooded too. So are turtles. The body temperature of a warmblooded animal stays about the same all the time. A warmblooded animal's temperature does not change when the temperature of their surroundings changes. What are some warmblooded animals? Birds are warmblooded. Squirrels and cats are other examples of warmblooded animals.

1. The body temperature of a coldblooded animal _____ when the
   changes, stays the same

   temperature of its surroundings changes.

2. The body temperature of a warmblooded animal _____ when the
   changes, stays the same

   temperature of its surroundings changes.

3. Fish and frogs are _____ animals.
   warmblooded, coldblooded

4. Birds are _____ animals.
   warmblooded, coldblooded

5. Turtles are _____ animals.
   warmblooded, coldblooded

People are warmblooded. However, people could not survive extreme cold or heat. So people often adapt the environment to meet their needs. For example, in very cold weather people heat their homes and wear warm clothing. In very warm weather, people wear lighter clothing.

Figure G

6. Why should you wear several layers of clothing in the winter?

   _____

7. How do people control the indoor environment during warm weather?

   _____

FILL IN THE BLANK

Complete each statement using a term or terms from the list below. Write your answers in the spaces provided. Some words may be used more than once.

environment	water	roots
oxygen	carbon dioxide	outside
repair	food	

1. Everything that surrounds an organism is called its _____ .

2. The things that living things need to stay alive are proper temperature,

    _____ , _____ , and _____ .

3. Living things get what they need from the _____ .

4. To make their own food, plants need the gas _____ .

5. Living things are made up mostly of _____ .

6. Water is taken in by a plant through its _____ .

7. Food provides organisms with the materials needed for growth and

    _____ .

8. Fish take in oxygen that is dissolved in the _____ .

9. During respiration, oxygen combines with digested _____ to produce

    energy.

10. Animals take in their food from the _____ .

REACHING OUT

Snakes are coldblooded animals. Why do you think snakes stay in the shade during the

hottest part of a summer day? _____

_____

_____

Lesson **7**

Where do living things come from?

KEY TERM

spontaneous generation: idea that living things can come from nonliving things

LESSON 7 | Where do living things come from?

At one time, people believed that living things could come from non-living matter. For example, people believed that worms and flies grew from rotting meat and that mice came from straw. The idea that living things can come from nonliving things is called **spontaneous generation** [spahn-TAY-nee-us jen-uh-RAY-shun].

Francesco Redi was an Italian scientist who lived during the 1600s. Redi did not think that maggots, or newly hatched flies, came from meat. He thought that living things could come only from other living things. Redi did an experiment to prove this. A description of Redi's experiment follows:

Redi knew that maggots are often found on decaying meat. He also knew that flies were attracted to the smell of the decaying meat. Redi put some spoiled meat into each of three jars. He left one jar uncovered. He put a thin net over a second jar. He sealed a third jar tightly with a lid. You can see the set up for Redi's experiment on page 39.

After a few days, Redi saw that maggots grew only on the meat in the uncovered jar. Why did the maggots grow there? The maggots hatched from eggs that flies had laid on the meat. No maggots appeared on the meat the flies could not reach. Redi showed that maggots did not come from the meat. He helped prove that living things come only from other living things of the same kind.

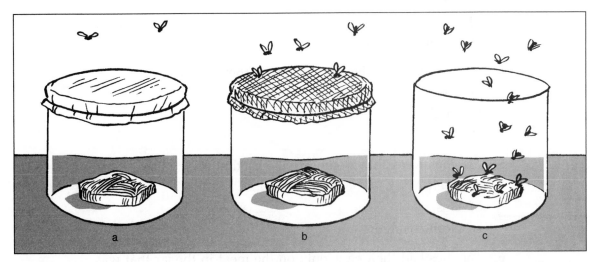

One jar was capped airtight.   *One jar was covered only with cheesecloth.*   *One jar was left open.*

Figure A

What Redi Saw

1. Flies hardly ever flew to the airtight jar.

2. Flies flew to the cloth-covered jar, but they could not reach the meat.

3. Flies flew into the open jar. They laid their eggs there.

4. Maggots developed <u>only</u> in the open jar.

What Redi Reasoned and Concluded

See if you can reason like Redi did. In the space provided, write the letter of the word or words that best completes each statement.

_____   1. The flies <u>could</u> smell the meat that was
   a) in the open container only.
   b) in the open container and in the cloth-covered container.
   c) in the airtight container.

_____   2. The flies <u>could not</u> smell the meat that was
   a) in the open container.
   b) in the open container and in the cloth-covered container.
   c) in the airtight container.

_____ **3.** Flies rarely flew to the airtight jar because the flies could not
  a) see the meat.
  b) feel the meat.
  c) smell the meat.

_____ **4.** The flies went to the jars with meat they could
  a) see.
  b) smell.
  c) feel.

_____ **5.** The flies were able to get into the jar that was
  a) open.
  b) closed.
  c) cloth-covered.

_____ **6.** The flies laid their eggs only on the meat in the jar that was
  a) open.
  b) closed.
  c) cloth-covered.

_____ **7.** Maggots developed only where
  a) the meat was decaying.
  b) the flies had laid their eggs.
  c) the flies had not laid their eggs.

_____ **8.** The maggots came from the
  a) meat.
  b) eggs.
  c) jar.

_____ **9.** When the eggs hatched they became
  a) maggots.
  b) flies.

_____ **10.** Life comes <u>only</u> from matter that is
  a) dead.
  b) living.
  c) not living.

What is a cell?

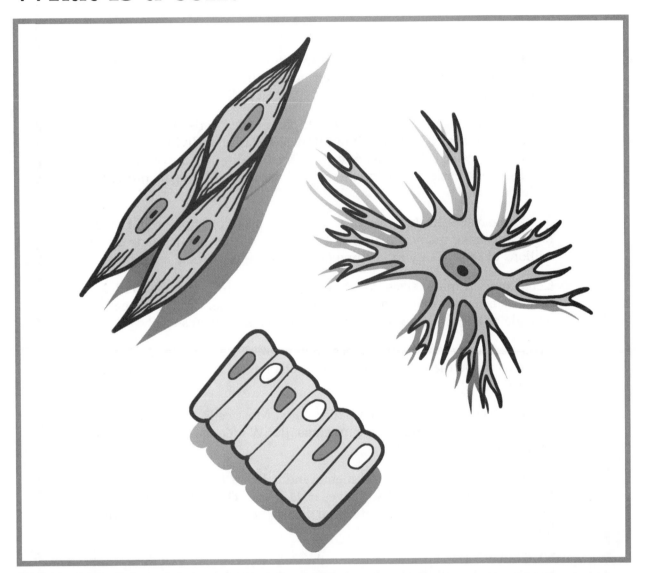

KEY TERMS

cell: basic unit of structure and function in all living things

cell membrane: thin "skin" that covers the cell and gives the cell its shape

nucleus: part of a cell that controls the cell's activities

cytoplasm: living material inside the cell membrane, excluding the nucleus

LESSON 8 | What is a cell?

The Great Pyramids of Egypt are made of stone blocks. Buildings are put together with bricks. Birds build their nests with grass and twigs. Everything is made up of smaller parts . . . EVEN YOU!

All living things are made up of small parts called **cells**. The cell is the basic unit of structure in all living things. Because all living things are made up of cells, cells often are called "the building blocks of life." The cell also is the basic unit of function in living things. All the life processes are carried out by cells.

Some organisms, such as bacteria, are made up of only one cell. Larger organisms have many more cells. A person, for example, is made up of trillions of cells. Can you imagine how many cells a whale must have?

Cells come in many sizes. Most are microscopic [my-kruh-SKAHP-ik]. Some cells, however, can be seen easily. For example, a chicken's egg is a single cell. Do you need a microscope to see a chicken's egg?

Cells also come in many shapes. For example, a muscle cell has a different shape than a nerve cell. Skin cells have a different shape than fat cells.

The cell itself is made up of smaller parts. Most cells have three main parts: the **cell membrane**, the **nucleus** [NEW-klee-us], and the **cytoplasm** [SYT-uh-plaz-um].

CELL MEMBRANE The cell membrane is like a thin skin that covers the cell. It protects the cell and gives it its shape. The cell membrane has tiny holes in it. Materials enter and leave the cell through these tiny holes.

NUCLEUS The nucleus is inside the cell. It controls everything that happens in the cell. The nucleus is like the "boss" of the cell. The nucleus usually is near the center of a cell.

CYTOPLASM The cytoplasm is the material located between the nucleus and the cell membrane. It fills most of the inside of the cell and contains many small structures. Like the cell membrane, the cytoplasm helps give a cell its shape. Most life functions take place in the cytoplasm.

Figure A shows eight cell parts. These parts are found in most cells. The name of each cell part is listed below.

- **cell membrane**
- **cytoplasm**
- **nucleus**
- **nuclear membrane**

- **mitochondria** [myt-uh-KAHN-dree-uh]
- **ribosomes** [RY-buh-sohmz]
- **endoplasmic reticulum** [EN-duh-plaz-mic rih-TIK-yuh-lum]
- **vacuoles** [VAK-yoo-wohls]

Each cell part is described below the diagram. As you read each description, identify the cell part in the diagram.

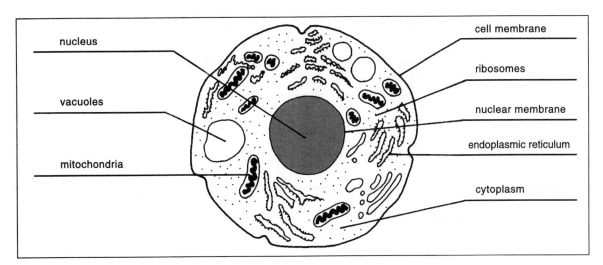

Figure A *A typical animal cell.*

CELL MEMBRANE

A thin covering that surrounds the cell. The cell membrane.

a) protects the cell,

b) helps give the cell its shape,

c) allows materials to enter and leave the cell, and

d) helps keep the cell material together.

CYTOPLASM

The living material inside the cell membrane but outside of the nucleus. Cytoplasm helps give a cell its shape. Most of the life functions take place within the cytoplasm.

NUCLEUS

A structure often found near the center of the cell. The nucleus is the "boss" of the cell. It controls all of the cell's activities. The nucleus is especially important during reproduction.

NUCLEAR MEMBRANE

A thin covering that surrounds the nucleus. The nuclear membrane controls the passage of materials into and out of the nucleus. It also gives the nucleus its shape.

MITOCHONDRIA Mitochondria are rod-shaped. They are the "power houses" of the cell. Mitochondria <u>store</u> and <u>release</u> the energy the cell needs to carry out the life functions.

ENDOPLASMIC RETICULUM A network of channels. The endoplasmic reticulum is like a series of "roadways." They are used for moving materials within the cell.

RIBOSOMES Tiny grainlike structures. The ribosomes make and store protein. Most ribosomes are found on the endoplasmic reticulum. Some, however, move freely within the cytoplasm.

VACUOLES Liquid-filled spaces. They store food and wastes. Some vacuoles also store extra water. They pump extra water out of the cell.

Answer the following questions about cells.

1. Where do most of the life functions take place within a cell? _____

2. Does each part of the cell work alone? Explain your answer. _____

   _____

   _____

3. What are two jobs of the cell membrane?_____

   _____

4. How are vacuoles like storage bins? _____

   _____

LABEL THE PARTS

Figure B shows an animal cell. Label each part of the cell on the lines provided.

1. _____

2. _____

3. _____

4. _____

5. _____

6. _____

Figure B

Plant cells and animal cells are not exactly alike. Plant cells have certain parts that animal cells do not. These parts are a cell wall and chloroplasts.

CELL WALL   The cell wall surrounds the cell membrane of a plant cell. The cell wall is made of a nonliving material called cellulose [SEL-yoo-lohs]. The cell wall is more rigid (stiff) than the cell membrane. It gives a plant cell its stiffness. It also gives it its shape.

CHLOROPLASTS   Chloroplasts are found in the cytoplasm of a plant cell. Chloroplasts contain a green substance called chlorophyll [KLAWR-uh-fil]. Chlorophyll is needed by green plants for food-making. The food-making process of green plants is called photo-synthesis [foht-uh-SIN-thuh-sis]. Most chlorophyll is found in the leaf cells of green plants.

Plants can make their own food. Animals cannot. Animal cells do not contain chlorophyll.

The number and size of vacuoles also is different in plant and animal cells. Plant cells have only one or two vacuoles. The vacuoles are usually very large. Animal cells have many small vacuoles.

Figure C shows a plant cell. The parts of the plant cell that are shown in the figure include:

- cell membrane
- cytoplasm
- nucleus
- cell wall
- chloroplast
- vacuole

Find each part of the plant cell in Figure C.

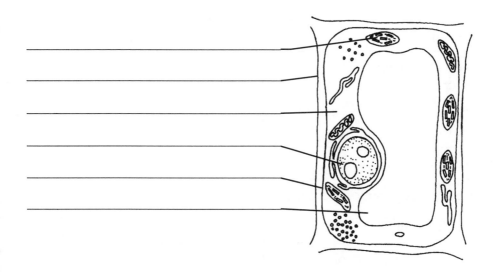

Figure C

Answer the following questions about plant and animal cells.

1. What two parts do plant cells have that animal cells do not have? _____

    and _____

2. What is the cell wall made of? _____

protoplasm, cellulose

3. Is cellulose living material? _____

yes, no

4. Where are the chloroplasts located? _____

in the nucleus, in the cytoplasm

5. What substance is found inside the chloroplasts? _____

protoplasm, chlorophyll

6. What is the substance inside the chloroplasts used for? _____

food-making, excretion

COMPLETE THE CHART

Answer the questions by putting a "YES" or "NO" in the space provided.

	Animal Cell	Plant Cell
1. Does it have a nucleus?		
2. Does it have ribosomes?		
3. Does it have mitochondria?		
4. Does it have a cell membrane?		
5. Does it have a cell wall?		
6. Does it have cytoplasm?		
7. Does it have chloroplasts?		
8. Does it have an endoplasmic reticulum?		
9. Does it have chlorophyll?		
10. Does it have many small vacuoles?		

THE "INS AND OUTS" OF THE CELL MEMBRANE

Substances must be able to get into and out of a cell in order for the cell to do its job. The passage of these materials takes place through the cell membrane by a process called diffusion [dih-FYOO-zhun]. In diffusion, some molecules may pass through tiny holes in the membrane. Others are carried across the membrane by special "carrier molecules." The molecules that are diffusing move to whichever side of the membrane has a lower concentration of that kind of molecule. For example, dissolved nutrients and oxygen tend to move into the cell. Dissolved wastes, such as carbon dioxide, tend to move out of the cell.

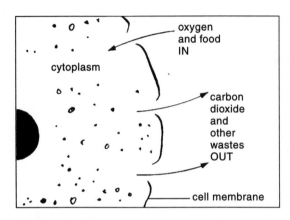

Figure D

The movement of water through a membrane is called osmosis [ahs-MOS-sis]. Osmosis is a special kind of diffusion. What might happen if the amount of water entering and leaving the cell were not controlled? It depends . . . The cell might

a) swell or

b) shrink

Think about each of the following possibilities. Write your answer in the space provided.

What might happen if:

1. too much water moved into a cell? _____

2. too much water moved out of a cell? _____

3. too little water moves into a cell? _____

4. too little water moves out of a cell? _____

5. water kept entering a cell and no water left the cell? _____

Answer the following questions.

6. What needed materials enter the cell through the cell membrane? _____

_____

7. What waste materials leave the cell through the cell membrane? _____

_____

FILL IN THE BLANK

Complete each statement using a term or terms from the list below. Write your answers in the spaces provided.

proteins   endoplasmic reticulum   cell wall
size   life processes   cell membrane
chlorophyll   nucleus   energy
cells   chloroplasts   shape

1. The "building blocks" of living things are _____ .

2. A plant cell has two parts that animal cells do not have. They are the

    _____ and the _____ .

3. The _____ is the control center of the cell.

4. Chloroplasts contain a green substance called _____ .

5. Ribosomes make and store _____ .

6. A network of channels called the _____ is used for moving
    materials within the cell.

7. A cell carries out all of the _____ .

8. The passage of substances into and out of a cell takes place through the

    _____ .

9. Cells vary in _____ and _____ .

10. Mitochondria store and release the _____ a cell needs to carry out the
    life processes.

REACHING OUT

*On a separate piece of paper, draw a picture of an animal cell. Try to do it from memory. Be sure
to include and label each of the parts listed below.*

- cell membrane   • cytoplasm   • nucleus   • nuclear membrane
- mitochondria   • ribosomes   • endoplasmic reticulum

Lesson 9

How can a microscope help us study living things?

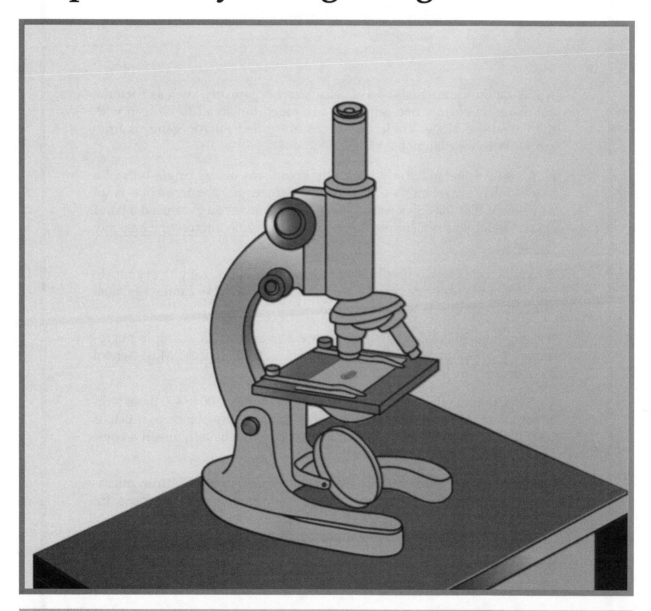

KEY TERM

microscope: instrument that makes things appear larger than they really are

49

LESSON 9 | How can a microscope help us study living things?

All living things are made up of cells. Some organisms, such as bacteria, are made up of only one cell. They are much too small to be seen with the naked eye alone. These organisms are called microorganisms [my-kroh-OWR-guh-nizums].

How can we see and study such tiny organisms or the single cells of a many-celled organism? We can use a **microscope**. A microscope is an instrument that makes objects appear larger. Have you ever used a hand lens? A hand lens is a simple microscope. A simple microscope has only one lens.

A hand lens is easy to use. It is small, and it does not weigh very much. But a hand lens does not magnify objects very much. We cannot see most one-celled organisms with a hand lens.

A compound microscope is much more powerful than a simple microscope. A compound microscope has two sets of lenses. Most school microscopes are compound microscopes.

Most compound microscopes can make objects appear 100 to 400 times larger than they really are. Some microscopes can magnify objects as much as 1000 times. When we talk about a microscope, we usually mean a compound microscope.

Another kind of microscope is the electron microscope. Electron microscopes can magnify objects up to 300,000 times. These microscopes are found in scientific laboratories.

Microscopes have many uses, especially in biology. Doctors often use microscopes. Have you ever seen a microscope in your doctor's office?

WHAT ARE THE PARTS OF A MICROSCOPE?

A compound microscope is shown in Figure A. The parts of the microscope have been labeled. Read the description of each part below the microscope. Then find the part in Figure A.

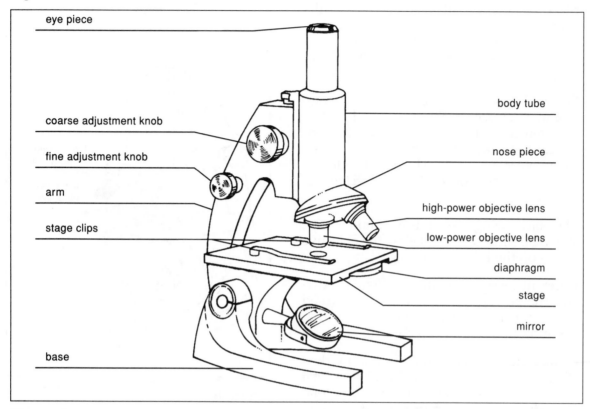

Figure A

Eyepiece or Ocular lens   Located at the top of the microscope. Holds the lens closest to the eye.

High-Power Objective   Longer of the two lenses close to the slide.

Low-Power Objective   Shorter of the two lenses close to the slide.

Body Tube   Gives the distance needed between the eyepiece and objective.

Coarse Adjustment Knob   Moves the tube up and down.

Fine Adjustment Knob   Moves the tube up and down, but only slightly.

Base   Holds up the entire microscope.

Arm   Supports the body tube.

Nosepiece   Holds objective lenses.

Mirror   Reflects light into the tube.

Diaphragm   Circular disk that adjusts the amount of light entering the stage area.

Stage   Platform that supports the slide; allows light to pass through.

Stage Clips   Hold the slide in place on the stage.

1.   What does the diaphragm do?

_____

_____

2.   What part of a compound microscope

supports the body tube? _____

3.   Which objective is longer?

_____

A microscope makes things look bigger. It does this because light coming from the object passes through lenses. A lens is a piece of glass that has been carefully shaped to bend light. Light that passes from an object through the lens of a microscope is bent so that the object looks larger.

Figure B shows the three lenses of a compound microscope. The top lens is called the eyepiece or the ocular lens. It is the lens closest to the eye. The other two lenses are called objective lenses. The objective lenses are the lenses closest to the object being viewed.

The object being viewed is on a microscope slide. The slide is placed below the objective lenses on the microscope stage.

Different lenses magnify to different powers. The power of magnification is marked by a number with an × next to it. A lens that magnifies ten times is marked 10×.

In this picture, the ocular lens is marked 10× and the objective 10×. This gives a total magnification of 100×. To find the total magnification of a microscope, just multiply the two magnifications.

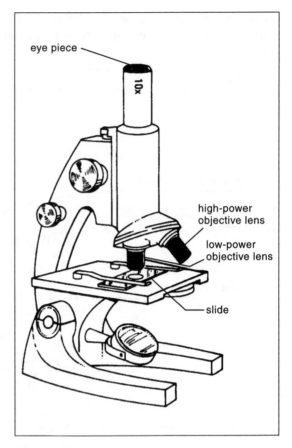

Figure B

Complete the table below by finding the total magnification for each pair of lenses. For example, the first pair has a total magnification of 100× (10 × 10 = 100).

Eyepiece	Objective lens	Magnification
10×	10×	100×
10×	40×	
10×	44×	
5×	10×	
5×	40×	
20×	10×	
20×	40×	

DISCOVERING MICROSCOPE FEATURES

Look at each picture. Then answer the questions next to the pictures.

Figure C

This is a picture of a common house fly. It has been magnified about two times.

1. Can you see much of the fly's detail?

   _____
   yes, no

Figure D

This is what a part of a fly looks like through a microscope. It is magnified 100 times.

2. You now see _____ of the fly,
   more, less

   but you see _____ detail.
   more, less

3. What part of the fly do you think this

   picture shows? _____

   _____

Figure E

This is the same part of the fly. This time it is magnified 400 times.

4. Compared to 100×, you now see

   _____ of the fly. However,
   more, less

   you see _____ detail.
   more, less

CONCLUSIONS

1. The higher the magnification of a microscope, the _____ of a specimen you see.
   more, less

2. The higher the magnification of a microscope, the _____ detail you see.
   more, less

Objects that are viewed with a microscope are placed on a small piece of glass called a microscope slide. A microscope slide is placed on the stage of a microscope below the objectives.

There are two kinds of microscope slides: temporary slides and permanent slides.

- A temporary slide cannot be stored. It is used once and then it is cleaned off. Temporary slides are useful for studying tiny organisms while they are still living.

- A permanent slide can be stored and studied over and over again. Only dead or nonliving things can be studied with a permanent slide.

Most microscope slides are made of glass and are about 7.5 cm × 2.5 cm (3"× 1"). Many times a cover slip is used on a slide. A cover slip is a very thin piece of glass or plastic, which rests on the glass slide.

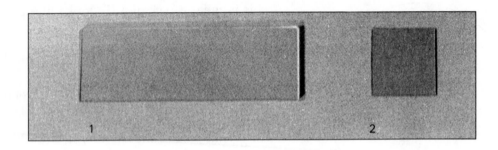

Figure F *Microscope slide and cover slip*

1. What type of slide would you use to study organisms living in pond water while

    they are still alive? _____

2. Why? _____

    _____

3. Where does a microscope slide get placed on a microscope? _____

4. What are the advantages of temporary slides? _____

    _____

5. What are the advantages of permanent slides? _____

    _____

Look at pond water under the microscope by following these directions.

1. Clean a microscope slide thoroughly.

2. Using a clean medicine dropper, place a drop of pond water in the center of the slide.

Figure G

3. Using forceps, gently lower a cover slip over the pond water.

4. Examine your slide under the microscope using the following procedure.

   a) First, rotate the nosepiece so that the low-power objective (10×) is in line with the body tube. You will hear a click when the objective is directly over the stage opening.

   b) Turn the coarse adjustment knob to find out which direction raises the objective and which direction lowers the objective.

Figure H

   c) Use the coarse adjustment knob to raise the low-power objective to about 2 cm above the stage.

   d) Look through the eyepiece. Be sure to keep both eyes open.

   e) Adjust the mirror.

   f) Move the diaphragm to adjust the amount of light.

g) Use the fine adjustment knob to focus clearly.

h) To view the object under high power, turn the nosepiece until the high-power objective clicks into place.

i) Look through the eyepiece and use the fine adjustment knob to focus under high power.

5. Draw a picture of what you see in the space below.

Some temporary slides do not need a cover slip.

To study salt or a piece of hair under the microscope, follow these directions.

1. Dampen the top surface of the slide very slightly with your finger.

Figure I

2. <u>Salt</u>—sprinkle a few grains on the glass.

   <u>Hair</u>—just place a piece of hair on the slide.

3. Now look at the specimens with the microscope.

4. Draw pictures of what you see in the space below.

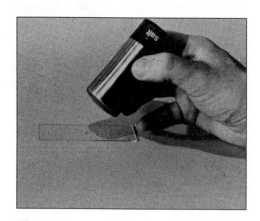

Figure J

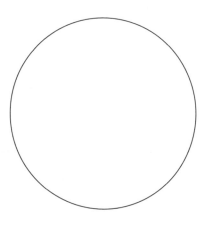

Salt

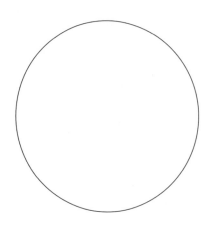

Hair

WHY DO SCIENTISTS USE STAINS?

Some things are transparent [trans-PER-unt]. You can see right through them. Glass is transparent. So are water and air. A transparent object lets light pass right through it.

Some cells and cell structures are transparent. This makes these structures hard to see under a microscope. How do scientists study such cells? They use a stain.

A stain is a dye. It adds color to the specimen on the slide. This makes the specimen easier to see.

There are many different kinds of stains. Not all cells and cell structures absorb, or take in, the same kind of stain. In addition, different parts of a cell may take in different amounts of stain. This means that some structures may be darker (or lighter) than others. The kind of stain used on a slide depends on what is being studied.

Look at Figure K. The photo shows blood cells as they appear under a microscope.

Scientists use a stain to make the white blood cells easier to see. Not all parts of the white blood cell absorb the same amount of stain.

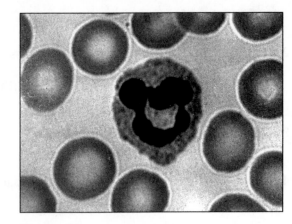

Figure K

1. What part of the blood cell has absorbed the most stain? _____

2. How do you know? _____

   _____

3. Why are stains used? _____

4. Name one transparent object. _____

   _____

5. What kind of cells are shown in Figure K? _____

   _____

   _____

Look at an onion leaf under the microscope by following these directions.

1. Separate an onion leaf from a quartered onion. Notice the very thin skin on the inner surface of the leaf. You might need to scrape the edge of the surface to find and separate this skin.

2. Remove a piece of the thin skin using a forceps.

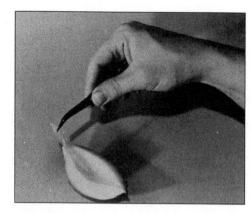

Figure L

3. Place a piece of skin on a microscope slide.

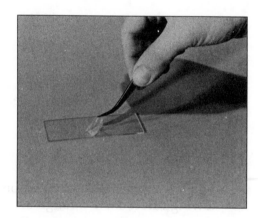

Figure M

4. Add a drop of iodine to the slide. Iodine is one kind of stain. The stain adds color to the onion skin. It makes the parts of the onion skin stand out more clearly.

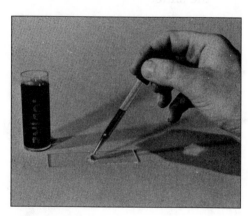

Figure N

5. Place a cover slip on the slide.

6. Examine your slide under the microscope.

Figure O

7. Draw a picture of what you see.

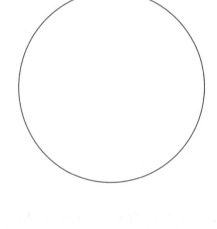

8. Prepare another slide of an onion leaf following directions 1 through 3.

9. Instead of iodine, place a drop of water on the slide.

10. Place a cover slip on the slide.

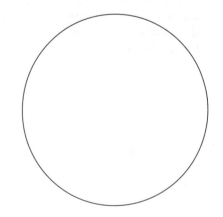

11. Examine your slide under the microscope.

12. Draw a picture of what you see.

OBSERVING A PERMANENT SLIDE

1. Obtain a permanent slide of human blood from your teacher.

2. Place the prepared slide of human blood on the stage of a compound microscope.

3. Use the low-power objective to focus on the slide. Examine the blood under the low power of your microscope. Then, switch to high power.

4. The blood on the prepared slide has been stained. Red blood cells will appear pink. The nuclei of white blood cells will appear blue-purple.

5. Find a red blood cell. Make a drawing of the cell in the space marked Red Blood Cell.

6. Still using the high power of the microscope, find a white blood cell. Make a drawing of it in the space marked White Blood Cell.

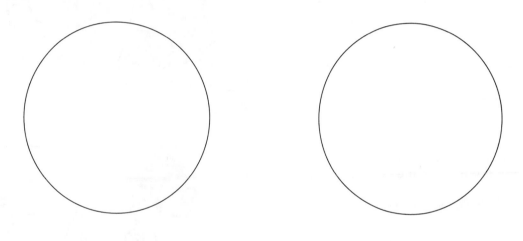

Red Blood Cell   **White Blood Cell**

COMPLETING SENTENCES

Choose the correct word or term for each statement. Write your choice in the spaces provided.

1. A mature red blood cell _____ a nucleus.

has, does not have

2. A white blood cell _____ a nucleus.

has, does not have

3. A permanent slide _____ be used again.

can, cannot

4. A temporary slide _____ be used again.

can, cannot

5. The blood in the above slide _____ stained.

was, was not

A microscope is a delicate instrument. Treat it carefully. A microscope should be held with two hands, one holding the arm of the microscope and the other supporting the base. The lenses of a microscope should only be cleaned with special lens paper. Regular tissues scratch the lenses.

Read about each picture. Then answer the questions next to each picture.

The student on the left is not holding the microscope the right way.

1. Can you describe the right way to

   hold a microscope? _____

   _____

   _____

   _____

   _____

Figure P

The boy is wondering which tissue to use to clean the microscope lenses.

2. Which one would you use?

   _____

3. Why? _____

   _____

   _____

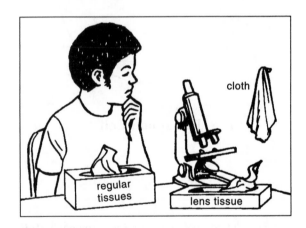

Figure Q

The girl is focusing downward toward the slide.

4. What has happened to the slide?

   _____

5. Is this the proper way to focus?

   _____

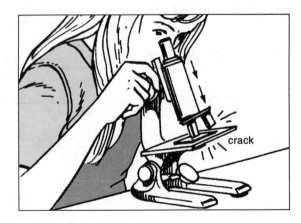

Figure R

6. What is happening to this boy's microscope? _____

_____

7. What should you do to prevent this from happening? _____

_____

_____

_____

Figure S

TRUE OR FALSE

In the space provided, write "true" if the sentence is true. Write "false" if the sentence is false.

_____ **1.** A microscope can have one lens.

_____ **2.** A transparent object blocks light.

_____ **3.** A compound microscope magnifies more than a simple microscope does.

_____ **4.** Light enters the eyepiece first.

_____ **5.** A microscope stage must have an opening.

_____ **6.** When you carry a microscope, you should hold it by the tube.

_____ **7.** A temporary slide cannot be stored.

_____ **8.** You should only use lens tissue to clean a microscope lens.

MATCHING

Match each term in Column A with its description in Column B. Write the correct letter in the space provided.

Column A

_____ **1.** simple microscope

_____ **2.** base

_____ **3.** compound microscope

_____ **4.** eyepiece

_____ **5.** transparent

Column B

a) supports entire microscope

b) has only one lens

c) allows light to pass through

d) has more than one lens

e) lens closest to the eye

FIND THE PARTS

Write the name of the microscope part next to the correct line in the picture. The parts are listed below.

arm	diaphragm	low-power objective
base	eyepiece	mirror
stage clips	fine adjustment knob	nosepiece
coarse adjustment knob	high-power objective	stage
body tube		

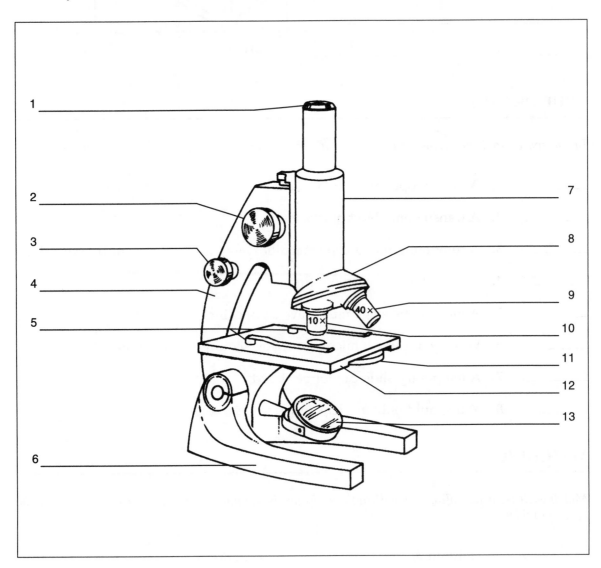

Figure T

What is reproduction?

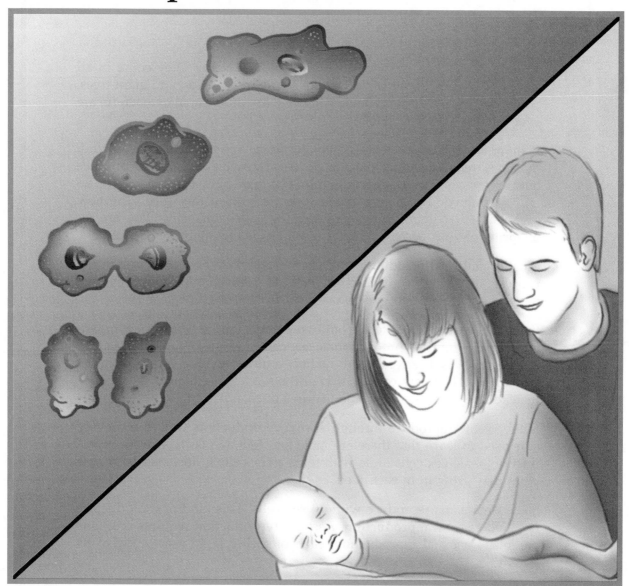

KEY TERMS

reproduction: life process by which organisms produce new organisms

sexual reproduction: a kind of reproduction that requires two parents where the cells from the two parents join to form a new organism

asexual reproduction: kind of reproduction that requires only one parent where the parent organism splits in half

LESSON 10 | What is reproduction?

Almost everywhere you look, you see living things. They come in all sizes and shapes. Some are large, like elephants, whales, and giant trees. Some, like ants, fleas, and blades of grass, are much smaller. Bacteria are even smaller—<u>MUCH</u> smaller!

Plant or animal, large or small, all living things <u>reproduce</u> their own kind. You may remember that the life process by which organisms produce new organisms is called **reproduction**. Having a baby is reproduction. So is the production of a new blade of grass from a tiny seed. The new organisms that living things produce are called <u>offspring</u>. You are the offspring of your parents. A litter of kittens is the offspring of a cat.

Most organisms (including people and cats) create new life through the process of **sexual reproduction**. Sexual reproduction requires two parents—one male and one female.

In sexual reproduction, cells from two parents join to form a new organism. Male organisms produce male sex cells. Female organisms produce female sex cells. In sexual reproduction, a new organism grows from the joined cells. The new organism is not exactly like either of its parents. Instead, the offspring has some features of each parent.

Many plants reproduce through sexual reproduction also. You may be surprised to learn that there are male and female plants. During reproduction, male sex cells unite with female sex cells to form seeds. A new plant may grow from each seed.

Some organisms reproduce without sex cells. This kind of reproduction is called **asexual reproduction**. Asexual reproduction requires only one parent. In asexual reproduction, each offspring is an exact copy of its parent.

Bacteria reproduce through asexual reproduction. A bacterium grows to full size. Then it divides in half. Two new bacteria are created.

Some plants also reproduce asexually. For example, strawberry plants grow runners. The runners root in the ground. A new plant then grows from the runner.

66

Read about each picture. Then answer the questions next to the picture.

Figure A shows a bacterium reproducing.

1. How do bacteria reproduce? _____

   _____

2. What will happen to the new bacteria?

   _____

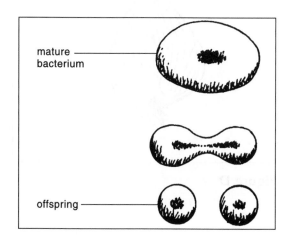

Figure A

In most animals, male and female sex cells are needed for reproduction. Only mature animals produce sex cells. Mature male animals produce <u>sperm</u>. Mature female animals produce <u>eggs</u>. Reproduction occurs when a sperm cell unites with an egg cell.

3. Can reproduction take place if the male or female does not mature?

   _____

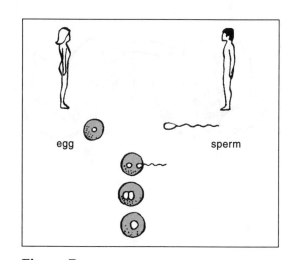

Figure B

A strawberry plant giving off runners is shown in Figure C.

4. How many runners are shown?

   _____

5. How many plants are shown?

   _____

6. Which plant is the oldest? _____

   _____

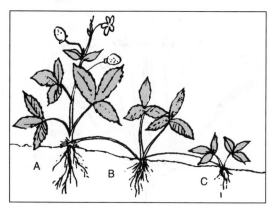

Figure C

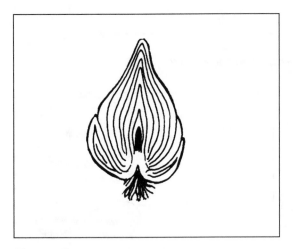

Figure D

Figure D shows the inside of a lily bulb. If a bulb is planted entirely beneath the ground, it produces a new plant. Some bulbs can be recognized by the layering.

7. Name a common bulb that you can probably find in your kitchen._____

   _____

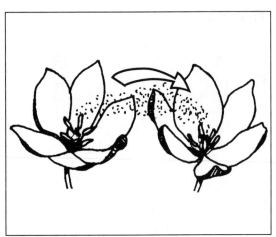

Figure E

The male reproductive cells of plants are called <u>pollen</u>. Plants reproduce sexually when pollen comes together with female reproductive cells, or eggs. Some flowers have only male cells or female cells. The union of pollen and egg cells produces seeds.

8. What will happen if the seeds are planted in the ground? _____

   _____

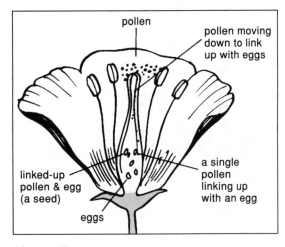

Figure F

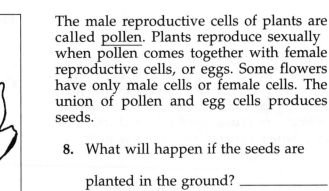

Some plants make both pollen cells and egg cells.

9. Why might this kind of reproduction be easier than having only male cells or female cells? _____

   _____

FILL IN THE BLANK

Complete each statement using a term or terms from the list below. Write your answers in the spaces provided.

dividing in half   parent   reproduction
organism   own kind

1. In sexual reproduction, male and female sex cells unite to form a new

    _____ .

2. All living things reproduce their _____ .

3. The process of producing new life is called _____ .

4. Asexual reproduction requires only one _____ .

5. Bacteria reproduce by _____ .

TRUE OR FALSE

In the space provided, write "true" if the sentence is true. Write "false" if the sentence is false.

_____ 1. Without reproduction, life could not continue.

_____ 2. Only mature organisms can reproduce.

_____ 3. A monkey can give birth to a cat.

_____ 4. Sexual reproduction requires two parents.

_____ 5. In asexual reproduction, each offspring is an exact copy of its parent.

WORD SCRAMBLE

Below are several scrambled words you have used in this Lesson. Unscramble the words and write your answers in the spaces provided.

1. EDES   _____

2. FFSOIPNGR   _____

3. EPRTNA   _____

4. BREATACI   _____

5. TOONIPERCRUD   _____

SCIENCE *EXTRA*

Hydroponics

Can you imagine growing plants in the snow-covered Arctic? Or how about on a ship at sea? Well, this may not be totally impossible, if hydroponics (hy-druh-PAHN-iks) is used.

What is hydroponics? Hydroponics is the science of growing plants without soil. Soil usually is needed by plants for support and nutrients. However, plants can live without soil if the plant gets support and nourishment from other sources. In addition, the plant must receive controlled amounts of carbon dioxide, oxygen, water, heat, and light, in order to carry out its life processes.

There are two methods of growing plants without soil. One method uses water. The plants are suspended with their roots submerged into water. The same nutrients found in fertile soil are dissolved in the water. Plants take in the nutrients from the water. A disadvantage to this method of hydroponics is that the plants must be supported from the top.

A second method of hydroponics uses materials other than water, such as sand or gravel. The plant's roots anchor it in the sand or gravel. A nutrient solution is sprayed onto the plants from above, or is pumped into the plant bed from below. Pumping nutrients into the plant bed is usually preferred because the roots of the plants grow downward and the nutrients get to the plant faster.

Although plants have been grown without soil since ancient times, hydroponics has been used commercially for less than 50 years. Some of the advantages of hydroponics include lack of competition from weeds, freedom from soil-transmitted diseases, and reduced labor costs. A disadvantage is the high cost of equipment needed for hydroponics.

What is mitosis?

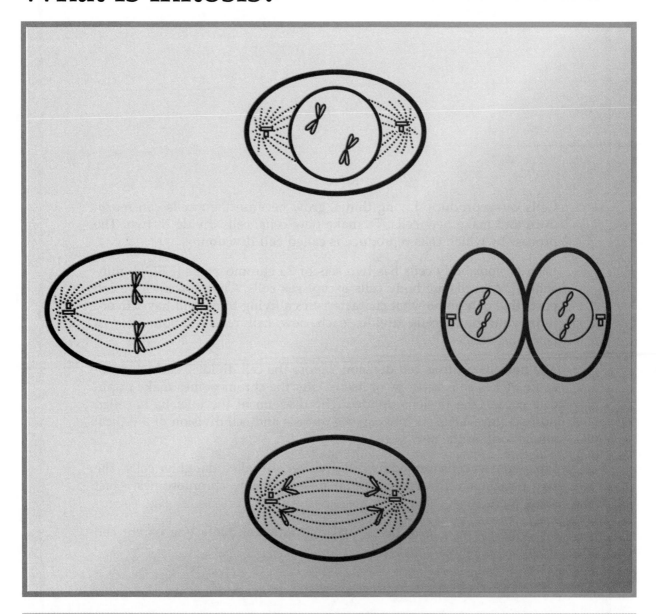

KEY TERMS

cell division: process by which cells reproduce

mitosis: kind of cell division in which the nucleus divides

LESSON 11 | What is mitosis?

Cells can reproduce. Living things grow because their cells can reproduce and make new cells. To make new cells, cells divide in two. The process by which cells reproduce is called **cell division**.

Each of your <u>body</u> cells has two sets of 23 chromosomes [KROH-muh-sohms]. All cells are body cells <u>except</u> sex cells. Chromosomes are cell parts that determine what characteristics a living thing will have. Every time your body cells divide, each new cell receives both sets of chromosomes.

The nucleus controls cell division. Before the cell divides, each chromosome makes an exact copy of itself. After the chromosomes make duplicate copies, the nucleus divides. The division of the nucleus is called **mitosis** [my-TOH-sis]. You can see mitosis and cell division of a typical animal cell on the next page.

The two new cells formed by cell division are called <u>daughter cells</u>. The two daughter cells are exactly alike—exact same chromosomes, exact same traits.

Cell division is a form of asexual reproduction. Only one parent cell divides to form daughter cells.

Let us study mitosis of a typical animal cell. This cell has two chromosomes. Follow what happens step-by-step.

Figure A

- The cell is getting ready to divide.

- The chromosome material is duplicating itself. It does not look like chromosomes—not yet.

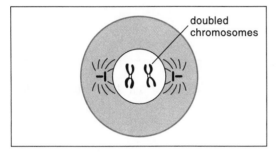

Figure B

- The chromosome pairs coil, becoming short and thick. They can be seen clearly. Count them.

- The nuclear membrane disappears.

- Spindle fibers form in the cytoplasm. The spindle fibers attach to the chromosomes and to both ends of the cell.

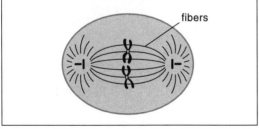

Figure C

- The two double chromosomes line up at the center of the cell.

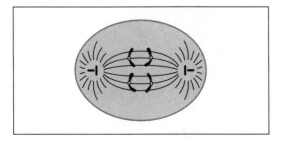

Figure D

- The halves of each doubled chromosome separate and move to opposite ends of the cell.

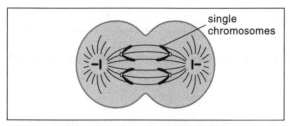

Figure E

- The cell begins to pinch in two.

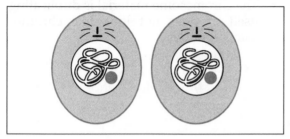

Figure F

- A new nuclear membrane forms in each daughter cell.

- The chromosomes uncoil.

- The cell has divided and there are now two daughter cells. Each daughter cell is exactly alike.

NOW TRY THIS

Study the diagrams below. In the spaces provided, number the diagram from 1 to 5 to show the correct order of the stages of mitosis.

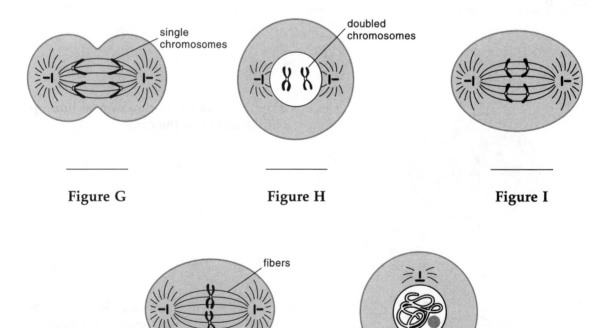

Figure G   **Figure H**   **Figure I**

Figure J   **Figure K**

Plant cells also reproduce by cell division. Like animal cells, plant cells make copies of themselves and carry out mitosis. However, in plant cells, a new cell wall and new cell membrane form down the middle of the cell. They form a wall between the two new nuclei. Two daughter cells are formed, one on each side of the new cell wall.

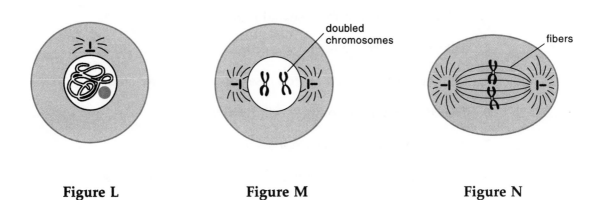

Figure L **Figure M** **Figure N**

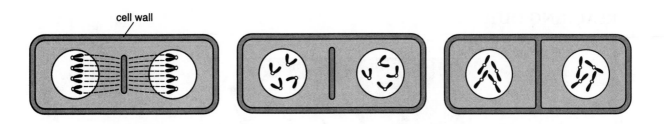

Figure O **Figure P** **Figure Q**

Using Figures L–Q, answer the following questions.

1. Is plant or animal cell division shown in the diagram? _____

2. How do you know? _____

3. What is happening in Figure N? _____

_____

4. What is happening in Figure O? _____

_____

5. What change occurs between the stages shown in Figure P and Figure Q?

_____

TRUE OR FALSE

In the space provided, write "true" if the sentence is true. Write "false" if the sentence is false.

_____ 1. Chromosomes determine what characteristics a living thing will have.

_____ 2. The division of the nucleus is called mitosis.

_____ 3. Cell division is a form of sexual reproduction.

_____ 4. Mitosis produces cells that are different from one another.

_____ 5. Daughter cells formed by cell division look exactly alike.

_____ 6. The cell membrane controls cell division.

_____ 7. All cells are body cells except sex cells.

_____ 8. Each of your body cells has two sets of 23 chromosomes.

_____ 9. Animal and plant cells divide in the same way.

_____ 10. Mitosis happens only in organisms that reproduce asexually.

REACHING OUT

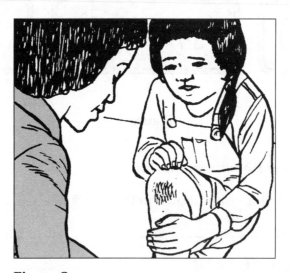

Figure S

Cells reproduce cells only of their own kind. From your own experience, how do you

know this is true? _____

_____

_____

What is binary fission?
What is budding?

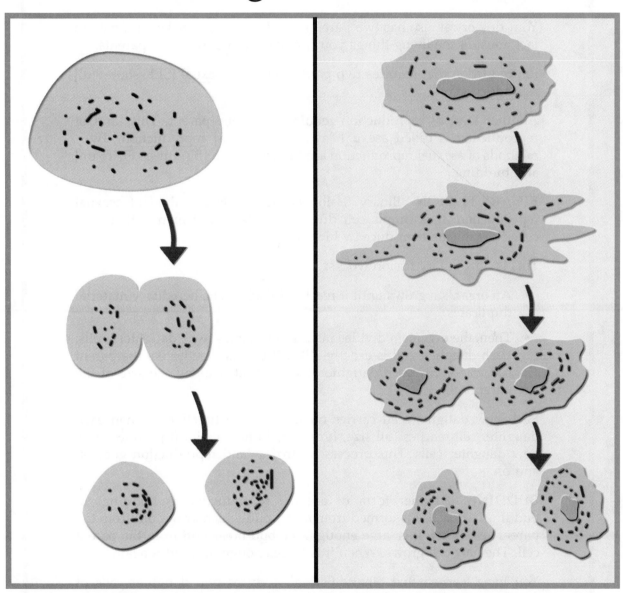

KEY TERMS

binary fission: form of asexual reproduction in which one cell divides into two identical cells

budding: form of asexual reproduction in which a small part of a cell breaks off to form a new organism

LESSON 12 | What is binary fission? What is budding?

You have a mother. You also have a father. You have two parents—one female and one male.

Your dog or cat also has two parents. So does a fly, a snake, or a fish. In fact, most of the living things you can name came from two parents.

Reproduction that requires two parents is called <u>sexual</u> [SEK-shoo-wul] <u>reproduction</u>.

Another kind of reproduction requires only one parent. This kind of reproduction is called <u>asexual</u> [ay-SEK-shoo-wul] <u>reproduction</u>. Two methods of asexual reproduction are **binary fission** [BY-nur-ee FIZH-un] and **budding**.

BINARY FISSION Binary fission is the simplest method of asexual reproduction. It is simple cell division. Bacteria and many other one-celled organisms reproduce by binary fission.

This is how binary fission works:

- An organism grows until it reaches full size. The hereditary material duplicates.

- Then, the organism divides in half. It becomes two "daughter" cells. Each daughter cell is exactly alike. It also is exactly like the parent cell was—only each daughter cell is about one-half the size of the parent cell.

Each new daughter cell carries on its own life functions. When each daughter cell reaches full size, it divides in half again. It produces two new daughter cells. This process of growth and reproduction goes on and on.

BUDDING Another form of asexual reproduction is budding. In budding, a new cell is formed from a tiny bud which grows out from the parent cell. When it is large enough, the bud breaks off from the parent cell. The new cell grows. When it is large enough, it divides again.

You have learned that binary fission produces two daughter cells of <u>equal</u> size. Budding is different. Budding produces two cells of <u>different</u> sizes. The cells are not exactly alike in size. The hereditary material of the offspring, however, is exactly the same as that of the parent cell.

Several kinds of organisms reproduce by budding. Yeasts are the most common. Yeasts are single-celled organisms.

UNDERSTANDING BINARY FISSION

Study Figures A, B, and C and answer the questions about each one.

Bacteria are microscopic organisms. They reproduce by binary fission.

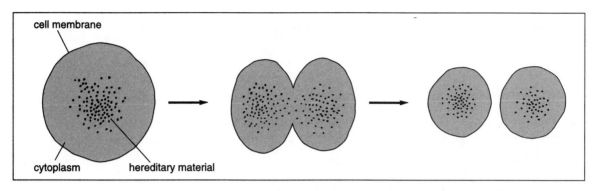

Figure A *A single bacterium [bak-TIR-ee-um] reproducing by binary fission*

1. What are the three parts of a bacterium shown? _____

    _____   _____

2. How does the nuclear material of the parent cell and each daughter cell compare?

    _____

    _____

3. About how much of the parent cell's cytoplasm does each daughter cell get?

    _____
    one-half, one-quarter, all of it

The paramecium [par-uh-MEE-see-um] and the amoeba [uh-MEE-buh] are simple one-celled organisms. They are microscopic in size. They reproduce by binary fission, too.

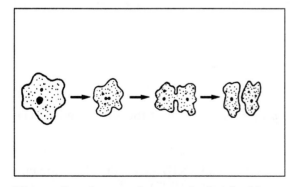

Figure B *An amoeba reproducing by binary fission*

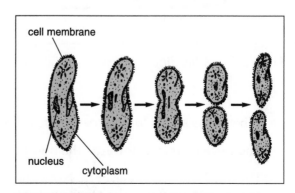

Figure C *A paramecium reproducing by binary fission*

4. What are the three cell parts shown in Figures B and C?

   _____   _____   _____

5. What happens to the nuclear material of the parent cell before it divides?

   _____

6. About how much of the parent cell's cytoplasm does each daughter cell get?

   _____

7. Are the daughter cells exactly alike? _____

8. Daughter cells are exactly like the parent cell, except in one way. What is it?

   _____

9. What kind of reproduction is binary fission? _____
                                                     sexual, asexual

10. How many parent cells take part in binary fission? _____

UNDERSTANDING BUDDING

Figure D shows a budding yeast cell. Read the explanation and then answer the questions.

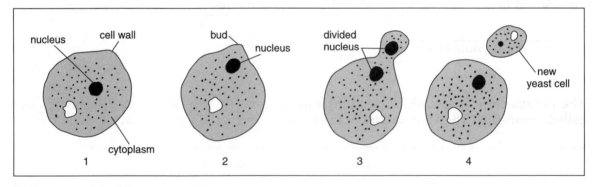

Figure D   *A budding yeast cell*

- When a yeast cell "buds," part of its cell wall bulges. This bulge is the start of a bud (1).

- The nucleus moves towards the bud (2).

- The nucleus divides evenly. Now there are two nuclei (NEW-klee-y). One nucleus moves into the bud. The other nucleus stays in the parent cell (3).

- The bud grows larger and larger. When it is large enough, it breaks off from the parent cell (4).

- The new yeast cell carries on the life functions. It takes in food and grows. When it is large enough, it reproduces by forming a bud, too.

1. In budding, the nucleus _____ .
   <u>divides evenly, divides unevenly</u>

2. One nucleus stays in the parent cell. The other moves _____ .
   <u>into the bud, out of the bud</u>

3. The amount of cytoplasm in the parent cell is _____ the amount in the bud.
   <u>more than, less than</u>

4. Budding is a kind of _____ reproduction.
   <u>asexual, sexual</u>

5. How many parent cells take part in budding? _____

FILL IN THE BLANK

Complete each statement using a term or terms from the list below. Write your answers in the spaces provided. Some words may be used more than once.

different	the same	reproduction
sexual	asexual	one-celled
binary fission	budding	

1. Creating new life is called _____ .

2. Reproduction that needs two parents is called _____ reproduction.

3. Reproduction that needs one parent is called _____ reproduction.

4. The simplest kind of asexual reproduction is called _____ .

5. Bacteria reproduce by _____ .

6. Amoeba are simple _____ organisms.

7. Amoeba reproduce by _____ .

8. Yeast cells reproduce by _____ .

9. Binary fission produces cells that are _____ size.

10. Budding produces cells of _____ sizes.

MATCHING

Match each term in Column A with its description in Column B. Write the correct letter in the space provided.

	Column A	Column B
_____	1. sexual reproduction	a) requires one parent
_____	2. asexual reproduction	b) produces cells of equal size
_____	3. binary fission	c) reproduce by budding
_____	4. budding	d) requires two parents
_____	5. yeasts	e) produces cells of different sizes

TRUE OR FALSE

In the space provided, write "true" if the sentence is true. Write "false" if the sentence is false.

_____ 1. Producing new life is called respiration.

_____ 2. Only animals reproduce.

_____ 3. Yeasts reproduce by binary fission.

_____ 4. Reproduction needing two parents is called sexual reproduction.

_____ 5. Reproduction needing only one parent is called asexual reproduction.

_____ 6. Bacteria and yeast reproduce by sexual reproduction.

_____ 7. Binary fission is a form of asexual reproduction.

_____ 8. Binary fission produces offspring of the same size.

_____ 9. Budding is a form of sexual reproduction.

_____ 10. Budding produces offspring of the same size.

REACHING OUT

1. When a bacterium reproduces, two daughter cells are produced. Under proper conditions, bacteria reproduce every 30 minutes.

   a) If you start out with one bacterium, how many bacteria will you have after one hour? _____

   b) How many will you have after two hours? _____

What are spores?

KEY TERMS

spores: reproductive cells

spore case: structure that contains spores

LESSON 13 | What are spores?

Have you sever seen "fuzz" growing on stale bread? If you have, you have probably seen bread mold. Mold is a many-celled organism. In some ways, molds are like yeasts. But molds do not reproduce by budding like yeasts do. Molds reproduce by special reproductive cells called **spores**.

The cells of a mold form many thread-like branches. Some of the threads are similar to roots. They grow down into the food the mold takes in.

Other thread-like branches grow straight up. At the top of each of these threads is a tiny ball. The ball, or **spore case**, contains thousands of spores. Spores are reproductive cells. A mold spore is a special cell that can reproduce other mold plants. (Remember, living things reproduce their own kind.)

When a spore case grows to full size, it bursts open. The spores fly into the air. They are very light in weight and are carried by the slightest air movement. Spores fall on everything. They are on you and everything around you.

Spores land on bread and other foods. If the temperature and moisture are right, the spores grow. They grow into new mold plants.

Other organisms, such as mushrooms and some plants, reproduce by spores. Reproduction by spores is still another form of asexual reproduction. It is the simplest kind of reproduction that uses special reproductive cells.

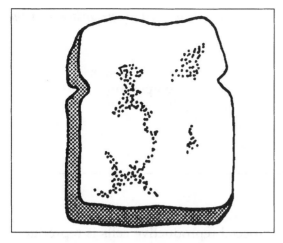

Figure A *Bread mold looks like this.*

Bread mold is fuzzy. At first it is white. Then it changes to gray and then to black. Mold often gives off a bad odor.

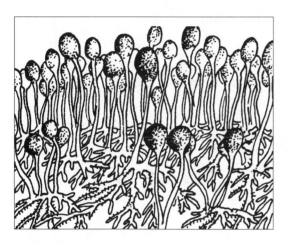

Figure B *This is what mold looks like under a microscope.*

Mold is thread-like. Some threads grow into the bread and take in food. The other threads grow straight up.

At the end of each upright thread is a spore case. A spore case holds thousands of special cells called spores. Each spore can reproduce into a new mold plant.

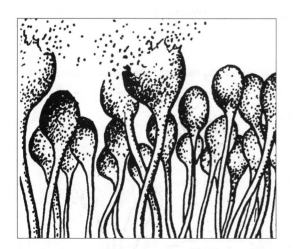

Figure C *When a spore case grows to full size it bursts open.*

The spores travel through the air. They land on everything.

The spores that land on food can reproduce into new mold plants. They reproduce if the temperature and moisture are right.

FILL IN THE BLANK

Complete each statement using a term or terms from the list below. Write your answers in the spaces provided. Some words may be used more than once.

budding	fuzzy	binary fission
mold	moisture	temperature
many	food	thread-like
spore case	spores	

1. Three kinds of asexual reproduction are _____ , _____ ,

   and reproduction by _____ .

2. The simplest kind of reproduction that uses special reproductive cells is reproduc-

   tion by _____ .

3. A mold is a simple _____-celled organism.

4. Without a microscope, a mold looks _____ .

5. Under a microscope we see that a mold is made up of many _____

   branches.

6. The thread-like branches of a mold that grow down take in _____ .

7. Each branch that grows upward has a _____ on the top.

8. A spore case contains thousands of _____ .

9. A single mold spore can reproduce into a new _____ .

10. A mold spore can reproduce into a new mold plant if it lands on _____ ,

    and if the _____ and _____ are right.

IS A SPORE A SEED?

A spore is not a seed. A seed is made by two parent cells—one male and one female. Seeds are produced through sexual reproduction.

A spore is made by one parent cell. Spores are produced by asexual reproduction.

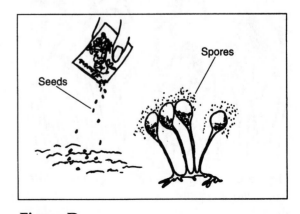

Figure D

What You Need (Materials)

a small piece of bread   small jar with cap
paper towel   water

Figure E

How To Do The Experiment (Procedure)

I. Prepare The Jar

1. Fold the paper towel in half, and in half again.

2. Cut a piece of the towel to fit the bottom of the jar.

3. Pour a small amount of water into the jar—just enough to wet the towel completely. Pour off the extra water.

II. Gathering The Spores

4. Wipe the bread across dust. (Every house has dust.) Try the top of a closet—the top of a door—any open place that you do not get to often.

5. Place the bread on the moist paper in the jar—dust side up.

6. Cap the jar <u>loosely</u>. Air must be able to get in. This is very important!

7. Place the jar in a dark place where it is not cold.

8. Look at it every day for a week.

Draw pictures in the boxes below showing how your mold looked as it was growing.

After 2 days   *After 4 days*

After 6 days   *After 8 days*

TRUE OR FALSE

In the space provided, write "true" if the sentence is true. Write "false" if the sentence is false.

_____   **1.** Reproduction by spores is a form of asexual reproduction.

_____   **2.** Molds reproduce by spores.

_____   **3.** A mold is a green one-celled animal.

_____   **4.** A mold makes it own food.

_____   **5.** A mold has many thread-like branches.

_____   **6.** Mold branches that grow downward feed the mold.

_____   **7.** A spore case contains seeds.

_____   **8.** Mold spores can reproduce yeast cells.

_____   **9.** Spores are very tiny.

_____   **10.** Every mold spore grows into a mold plant. (Think about this carefully!)

WORD SCRAMBLE

Below are several scrambled words you have used in this Lesson. Unscramble the words and write your answers in the spaces provided.

1. ROSEP   _____

2. DOLM   _____

3. ZYFUZ   _____

4. ARDEB   _____

5. UALSAEX   _____

REACHING OUT

1. Name three things that a spore needs to grow.

_____   _____   _____

2. Do you think a seed needs the same things to grow? _____

Lesson **14**

What is regeneration?

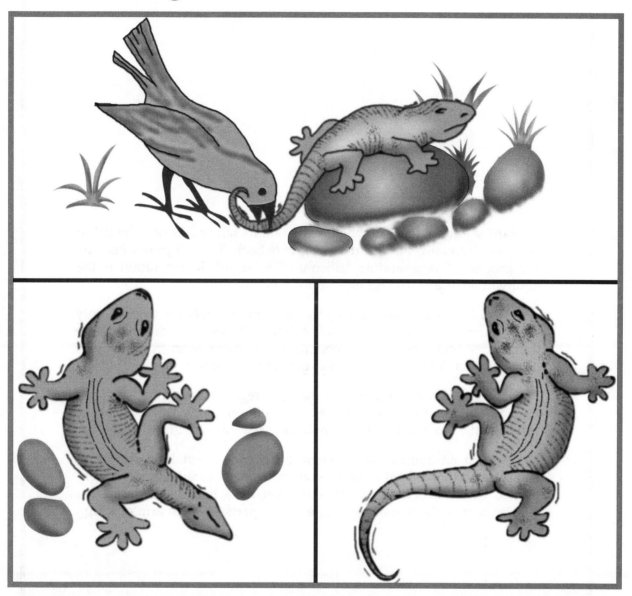

KEY TERM

regeneration: ability of an animal to regrow lost body parts

LESSON 14 | What is regeneration?

A lizard is attacked and grabbed by its tail. Its tail breaks off. The lizard escapes. Gradually, the lizard's tail grows back. The tail grows back by the process of **regeneration** [ri-jen-uh-RAY-shun]. Regeneration is the ability of an animal to regrow lost body parts.

Regeneration in animals varies greatly. Some animals can regrow only small parts. The lizard is one such animal. Others can regenerate large body parts. Still others can regenerate a whole organism from just part of the animal. This is a kind of asexual reproduction.

For example, a sea star can reproduce by regeneration. Most sea stars have five arms, or rays. If just one arm is cut off, along with part of the center of the sea star's body, a whole new sea star will grow.

How much an animal can regenerate depends upon how simple or complicated that animal is. Regeneration in complicated animals is very limited. Mammals can regenerate only skin, nails, hair, and certain other tissues. Mammals cannot regenerate whole parts—like an arm or a leg.

SEA STAR REGENERATION

Figures A and B show sea star regeneration. Study the figures and answer the questions.

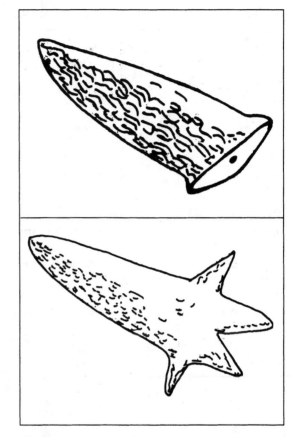

Figure A

A sea star can regenerate lost arms. However, a sea star can regenerate more than an arm.

Figure B

A sea star can regenerate a complete sea star from just one arm and part of the center.

1. **a)** Can more than one sea star regenerate from just one five-arm sea star?

    _____   **b)** How many? _____

2. **a)** Which figure shows a form of asexual reproduction? _____

A, B, both

    b) Explain your answer. _____

    _____

3. **a)** Which figure shows only regeneration of a body part? _____

A, B, both

    b) Explain your answer. _____

    _____

A *planarian* [pluh-NER-ee-un] is a tiny flatworm. It lives in ponds.

If a planarian is cut into pieces, it will regenerate its missing parts.

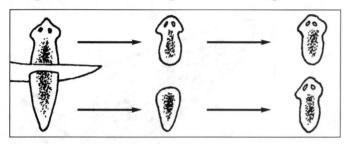

Figure C

A planarian can be cut in half across its body. The head portion will grow a new tail. The tail portion will grow a new head.

The way a planarian regenerates depends upon how it is separated. This happens in nature, but it can also be done in the laboratory.

Figure D shows four planaria. Each one was separated in a different way. (The dotted lines show the separations.)

Figure E shows four groups of regenerated planaria.

Which planarian regenerated into which group? Answer by placing the correct number on the line.

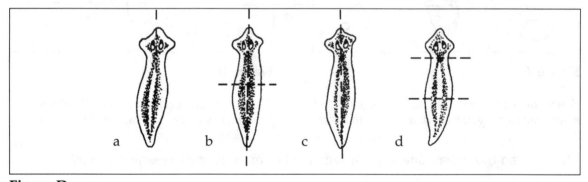

Figure D

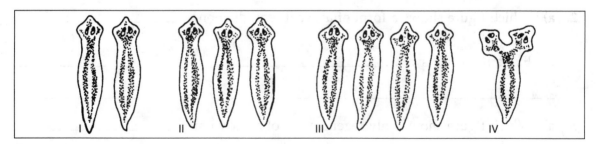

Figure E

1. Planarian **a** became group _____ . 3. Planarian **c** became group _____ .

2. Planarian **b** became group _____ . 4. Planarian **d** became group _____ .

Figure F

Every day billions of your body cells die.

Every time you wash your hands, you wash off hundreds—even thousands—of skin cells. New skin cells are always being regenerated.

Figure G
A deer sheds its antlers every year. New antlers grow back.

Figure H
If an earthworm is cut in two, the front half can grow into a complete worm if enough sections are left.

Figure I
The glass-tailed lizard escapes from its enemies by breaking off the end of its tail. Later, a new tail will grow.

Figure J
If a lobster or crab loses a claw to an enemy, it can grow a new one.

FILL IN THE BLANK

Complete each statement using a term or terms from the list below. Write your answers in the spaces provided. Some words may be used more than once.

nails	sea star	whole parts
simple	lizard	regeneration
lobster	skin	people
asexual reproduction	complicated	hair
five	dogs	planarian

1. Most sea stars have _____ arms.

2. The ability to regrow lost body parts is called _____ .

3. In animals, regeneration power greatly depends upon how _____ or

   how _____ the animal is.

4. Mammals are _____ animals.

5. Examples of mammals are _____ and _____ .

6. Mammals can regenerate tissues like _____ , _____ , and

   _____ .

7. Mammals cannot regenerate _____ .

8. Two simple animals that can regenerate a complete animal from just a part are the

   _____ and the _____ .

9. Animals like the _____ and the _____ can grow certain
   whole parts. But they cannot regenerate a whole organism from just a part.

10. Regeneration that forms a whole organism from just one part is said to be a kind of

    _____ .

REACHING OUT

Sea stars feed on oysters. At one time, oyster fishermen tried to kill the sea stars. They scooped up the sea stars from the oyster beds. They chopped them up and dumped the pieces back into the water. What do you think happened? Finish the story in your own words.

_____

_____

_____

What is vegetative propagation?

KEY TERMS

vegetative propagation: asexual reproduction in plants

tuber: underground stem

bulb: underground stem with fleshy leaves

cutting: part of a plant that is removed to grow another plant

grafting: attaching a cutting of one plant to another plant

LESSON 15 | What is vegetative propagation?

Plants are living things. Therefore, they reproduce. Many plants reproduce from seeds. Others reproduce without seeds. Reproduction from seeds is sexual reproduction. Reproduction without seeds is asexual reproduction. Asexual reproduction in plants is called **vegetative propagation** [VEJ-uh-tayt-iv prahp-uh-GAY-shun]. In vegetative propagation, roots, stems, or leaves reproduce one or more new plants. There are several kinds of vegetative propagation. Some kinds are natural. Other kinds are artificial.

Two kinds of natural vegetative propagation take place by means of **bulbs** and **tubers**.

BULBS   A bulb is really a short underground stem surrounded by thick colorless leaves. They are special leaves. They do not make food like green leaves do. They store the food made by the green leaves above the ground. The leaves protect and nourish the bulb.

The bulb you know best is the onion. An onion plant produces many bulbs. Each bulb may grow into a new onion plant. Other plants that grow from bulbs are tulips and lilies.

TUBERS   A tuber also is a heavy underground stem. It stores food made by the green leaves above the ground. The white potato is a tuber. (The sweet potato is not a tuber.)

A tuber has several "eyes." Each eye is really a bud. Each bud can sprout into a complete new plant.

Two kinds of artificial propagation take place by means of **cuttings** and **grafting**.

CUTTINGS   You may have made a cutting from a house plant yourself. You cut the stem or a leaf from a plant. You placed the cutting in water or in moist soil or sand. In a few days, roots developed. Then you could plant the cutting in soil.

Geraniums may be grown from stem cuttings. Begonias may be grown from leaf cuttings.

GRAFTING   Attaching a stem cutting of one plant to another plant is called grafting. The plants then grow together. A plant can be grafted only to another plant that is related.

Grafting is most often done with fruit trees.

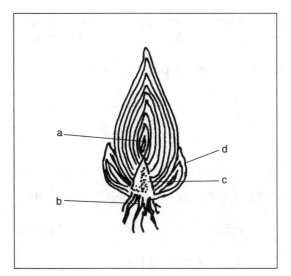

Figure A

Figure A shows a bulb. It has colorless leaves (a), roots (b), a stem (c) and bulblets (d). Look at an onion and see if you can identify these parts.

1. What is the job of the colorless leaves?

   _____

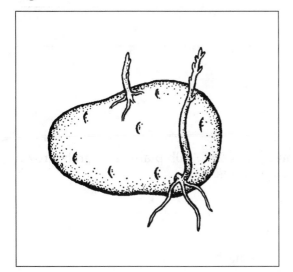

Figure B

Figure B shows a tuber. Notice the "eyes" of the potato. Also look for the smaller roots growing out of the "stem."

2. What is the job of a tuber? _____

   _____

   _____

   _____

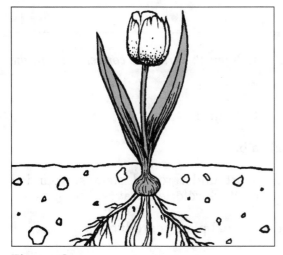

Figure C

Figure C shows what a complete bulb plant looks like.

It has **roots, above-ground stems, green leaves, colorless leaves,** and a **flower**.

3. Which part takes in water and

   minerals? _____

4. Which part of the plant makes food?

   _____

5. Where is this food stored? _____

   _____

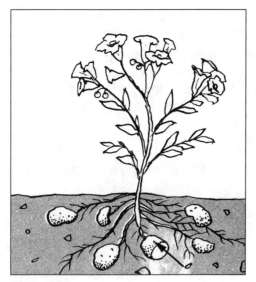

Figure D

This is what a complete potato plant looks like.

It has **roots, tubers, above-ground stems, green leaves,** and several small **flowers**.

6. Where does a potato plant make its food? _____

7. Where is this food stored? _____

_____

8. Find the part of a tuber labeled with an arrow—one of the potato's "eyes." What was it used for?

_____

SOMETHING INTERESTING

- Flowers produce fruits.

- Fruits have seeds.

- Seeds produce new plants.

- Bulb and tuber plants have flowers, fruits, and seeds. Yet, bulb plants are rarely grown from seeds. Potato plants are never grown from seeds!

WHY?

- The seeds of bulb plants take too long to grow.

- Potato seeds are weak. They rarely grow into plants.

MATCHING

Match each term in Column A with its description in Column B. Write the correct letter in the space provided.

Column A

_____ 1. vegetative propagation

_____ 2. tuber

_____ 3. tuber eye

_____ 4. bulb

_____ 5. tubers and colorless leaves of bulbs

Column B

a) a bud

b) short underground stem surrounded by thick colorless leaves

c) asexual plant reproduction

d) store food

e) heavy underground stem

REPRODUCTION FROM CUTTINGS

Figure E *A stem cutting*

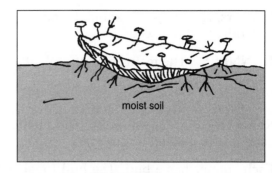

Figure F *A leaf cutting*

Some plants that reproduce from stem cuttings are geraniums, ivy, and many types of grapes. Roses also can be reproduced from stem cuttings.

African violets and some begonias reproduce from leaf cuttings.

Reproduction from cuttings is _____ vegetative propagation.
<u>natural, artificial</u>

REPRODUCTION BY GRAFTING

Figure G

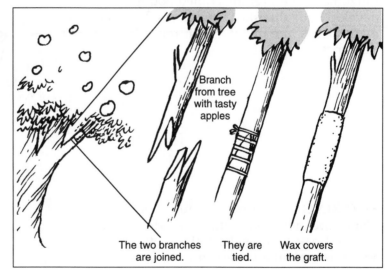

Branch from tree with tasty apples

The two branches are joined.   They are tied.   Wax covers the graft.

Figure H

The large tree in Figure G is very strong and can live in bad conditions. But the apples from it are small and do not taste good.

Branches from apple trees that produce large and tasty apples may be grafted to the strong tree. Figure H shows how it is done.

What kind of apples will the grafted branches produce? _____

Runners, rhizomes [RY-zohmz], and layering are three more ways that plants reproduce naturally without seeds.

RUNNERS The main stem of a plant grows straight up. It gives the plant support. Some plants have other kinds of stems, too. They are special reproductive stems called runners. Runners grow outward from the plants and close to the ground (Figure I).

Each runner has a bud. The bud touches the earth and starts a new plant. Strawberry plants, for example, reproduce by runners.

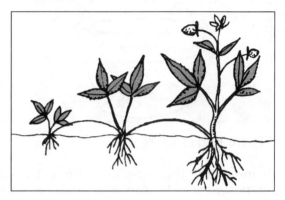

Figure I *A strawberry plant reproduces by runners.*

RHIZOMES A rhizome is a thick, underground stem (Figure J). It contains stored food. Rhizomes grow outward from a plant.

Rhizomes have swellings called nodes. Nodes develop buds that start new plants. Irises and many ferns are plants that reproduce by rhizomes.

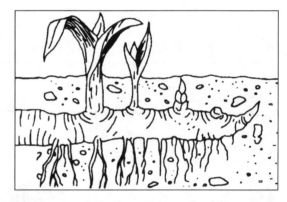

Figure J *A fern reproduces by rhizomes.*

LAYERING The upright stems of certain plants are not very stiff. They droop. If a drooped part touches the earth, roots develop and a new plant grows (Figure K).

Layering happens in nature. It can also be done artificially. Rose, raspberry, and blackberry plants are examples of plants that reproduce by layering.

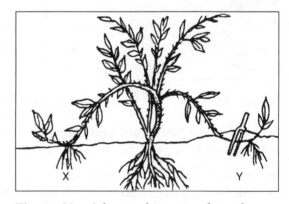

Figure K *A berry plant reproduces by layering. Reproduction by layering can be natural or artificial.*

WHAT DO THE PICTURES SHOW?

Look at Figures I, J, and K and answer the questions about each.

1. How many runners do you see in Figure I? _____

2. In Figure I, the oldest plant is on the _____ .
   _{right, left}

3. The youngest plant is on the _____ .
   _{right, left}

4. Runners grow _____ the ground.
   _{above, under}

5. A runner is a special kind of _____ .
   _{root, stem, leaf}

6. What type of plants reproduce by rhizomes?

   _____   _____

7. A rhizome is a special kind of underground _____ .
   _{root, stem, leaf}

8. In Figure K, which side do you think shows natural layering? _____
   _{X, Y}

9. Which side shows artificial layering? _____
   _{X, Y}

10. A berry plant has _____ stems.
    _{stiff, droopy}

DO THIS AT HOME

Vegetative propagation is the asexual reproduction of a whole plant from a plant part. Grow your own whole plants from plant parts. Use Figure L as a guide. It shows you what to do. Be patient—your new plants will take time to grow.

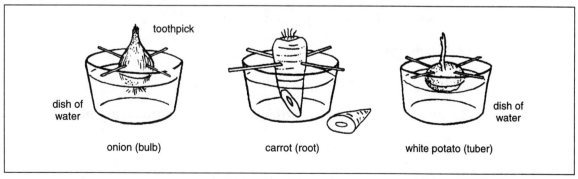

Figure L

FILL IN THE BLANK

Complete each statement using a term or terms from the list below. Write your answers in the spaces provided. Some words may be used more than once.

nodes   moist   underground stem
vegetative propagation eyes   layering
related   colorless leaves   outward

1. Asexual reproduction in plants is called _____ .

2. Rhizomes have swellings called _____ .

3. A bulb is a short _____ .

4. A bulb is surrounded by thick _____ .

5. Runners grow _____ from a plant.

6. Drooping plant stems may cause a plant to reproduce by _____ .

7. Cuttings must be kept _____ .

8. A tuber is a thick _____ .

9. New plants grow from the _____ of a parent tuber.

10. Grafting is done only with _____ plants.

TRUE OR FALSE

In the space provided, write "true" if the sentence is true. Write "false" if the sentence is false.

_____ **1.** Plants reproduce asexually from runners and rhizomes.

_____ **2.** A potato is a bulb.

_____ **3.** Rhizomes grow under the ground.

_____ **4.** A tuber is a short underground flower.

_____ **5.** All stems grow upward.

_____ **6.** Stiff plants reproduce by layering.

_____ **7.** Cuttings and grafting are artificial means of reproducing plants.

_____ **8.** A cutting may be a stem or a leaf.

_____ **9.** A stem or leaf cutting should be kept dry.

_____ **10.** All kinds of plants can be grafted together.

How do animals reproduce?

KEY TERMS

fertilization: union of one sperm cell and an egg cell

zygote: fertilized egg

LESSON 16 | How do animals reproduce?

Most animals reproduce sexually. You have learned that for sexual reproduction to take place, two parents are needed. The male provides the male sex cells, or sperm; the female provides the female sex cells, or eggs. Sex cells also are called gametes.

In some animals, special organs of the body are needed for reproduction. Male organs, called testes [TES-teez], produce sperm. Female organs, called ovaries [OH-vur-eez], produce eggs.

The process of sexual reproduction starts when a sperm cell and an egg cell unite. **Fertilization** [fur-tul-i-ZAY-shun] is the union of one sperm cell and an egg.

In some animals, fertilization takes place outside the body. In other animals fertilization takes place inside the female's body.

Fertilization that takes place outside the body is called external fertilization. Goldfish and frogs are two animals that reproduce by external fertilization.

Fertilization that takes place inside the body is called internal fertilization. Birds, snakes, and dogs are some animals that reproduce by internal fertilization.

The fertilized egg is called a **zygote** [ZY-goht]. A zygote is the beginning of a new life. A zygote is a single cell. Soon after it forms, the zygote divides. It becomes two cells. Each of these cells then divides. The two cells become four cells. Then these cells divide again. Cell division continues over and over again. A young organism, or embryo, forms.

As the cells divide, they form tissues and organs. The embryo grows in size. When the embryo is completely developed, birth takes place. The offspring is now an organism on its own. It must carry out all the life functions by itself.

EGG AND SPERM CELLS ARE DIFFERENT

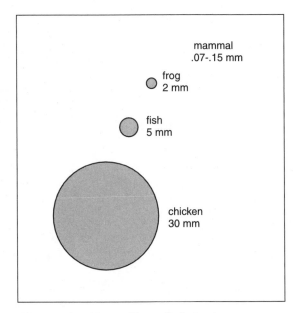

Figure A *Egg cells and their sizes*

An egg is round and large. Some eggs can be seen with the eye alone. An egg cannot move itself from place to place.

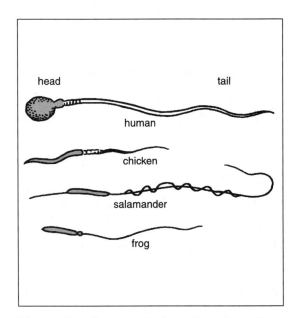

Figure B *Sperm cells (greatly enlarged)*

A sperm is free-swimming. A sperm has a "head" and a "tail." The tail lashes back and forth. This moves the sperm forward toward the egg.

FERTILIZATION

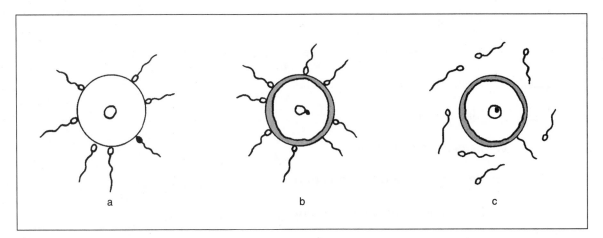

Figure C *Fertilization of a human egg*

Study Figure C. Then answer the questions.

1. Which is larger, a sperm or an egg? _____

2. A sperm is _____ smaller than an egg.

slightly, much

3. Which is the male gamete? _____

4. Which is the female gamete? _____

5. Which one is free-moving? _____

6. How many sperm swim toward an egg? _____

only one, many

7. How many sperm enter the egg? _____

only one, many

8. How many sperm fertilize the egg? _____

only one, many

9. Which part of the sperm enters the egg? _____

10. Which part of the sperm is left behind? _____

TRUE OR FALSE

In the space provided, write "true" if the sentence is true. Write "false" if the sentence is false.

_____ **1.** Only animals reproduce sexually.

_____ **2.** All animals reproduce sexually.

_____ **3.** Sexual reproduction needs two parents.

_____ **4.** Females produce egg gametes.

_____ **5.** Males produce sperm gametes.

_____ **6.** Eggs are free-moving.

_____ **7.** Egg cells are larger than sperm.

_____ **8.** Many sperm cells fertilize one egg.

_____ **9.** A fertilized egg divides many, many times.

_____ **10.** An embryo is a full-grown organism.

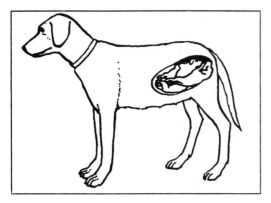

Figure D

Figure E

Animals like dogs, cats, horses, and whales are mammals. Humans are mammals, too. Mammal eggs are fertilized internally. The embryos develop internally too. When an embryo is fully developed, it is born. Female mammals produce milk to feed the newborn.

Figure F

Animals like birds and snakes are not mammals. The eggs of birds and snakes are fertilized internally. However, the female lays the fertilized eggs. The embryos then develop outside of the female's body. When the embryos are fully developed, they "hatch."

COMPLETE THE CHART

Answer the questions by putting a "YES" or "NO" in the space provided.

		Mammals	Birds	Snakes
1.	Is fertilization internal?			
2.	Is development internal?			
3.	Do females produce milk?			
4.	Do embryos hatch?			
5.	Is fertilization external?			

SOME INTERESTING FACTS ABOUT MAMMAL REPRODUCTION

The time between fertilization and birth is called the gestation [jes-TAY-shun] time. Gestation time varies greatly with different animals.

Animal (mammal)	Gestation time (approximate)
hamster	16½ days
house mouse	21 days
rabbit	30 days
dog or cat	63 days
lion	108 days
chimpanzee	237 days
human	267 days
cow	281 days
horse	336 days
elephant	660 days

Usually . . .

1. The smaller the animal, the

   _____ the gestation time.
   longer, shorter

2. The larger the animal the

   _____ the gestation time.
   longer, shorter

3. Which animal on the chart has the

   longest gestation time? _____

4. Which animal on the chart has the

   shortest gestation time? _____

5. Which has a longer gestation time,

   a human or a chimpanzee?

   _____

REACHING OUT

How many offspring do mammals usually produce at one time? It depends upon the animal. For example, a horse produces 1 offspring. An elephant also produces 1 offspring; usually, so does a human. Cats produce about 4 or 5 kittens and dogs produce from 1 to 12 puppies at a time. Lions produce 3 to 5 cubs. A mouse has 4 to 7 offspring.

What does the number of offspring tell us about the number of eggs fertilized?

_____

How many eggs does a human female usually release at one time? _____

How do you know? _____

How do fish reproduce and develop?

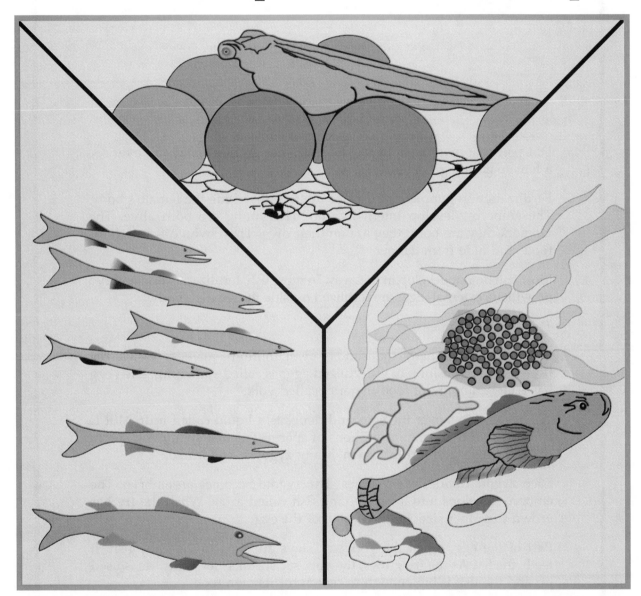

KEY TERMS

roe: fish eggs

spawning: fish breeding

milt: milky fluid of a fish that contains sperm

fry: young fish

yolk sac: part of the egg that feeds the fry

LESSON 17 | How do fish reproduce and develop?

Did you ever own a fish tank? Did you raise guppies? Did you ever see a female guppy giving birth?

Fertilization in guppies takes place internally—inside the female's body. The embryos develop internally, too. The young are born alive. The moment they are born, they are on their own. They swim off, search for food, and hide from danger.

Most fish do not develop in this way. In most fish, fertilization and embryo growth are external. Eggs are fertilized outside the female's body.

This is what happens:

The female lays many thousands—even millions—of eggs called **roe** [ROH]. The depositing of unfertilized eggs is called **spawning**. Each egg has a large amount of food material called yolk.

The male swims over the eggs and deposits a liquid called **milt**. Milt is a milky fluid that contains millions of sperm. The free-swimming sperm reach the eggs and fertilize them. Many zygotes are formed.

Each zygote divides many times. The zygote becomes an embryo. The embryo develops into a very young fish called a **fry**. When the fry has grown to proper size, it hatches out of the egg.

Part of the egg, called the yolk sac, stays attached to the baby fish. It feeds the fry. As the fry grows, the yolk sac shrinks. Soon, the sac is used up and the fish is big enough to search for its own food.

Study figures A through D. Answer the questions about each.

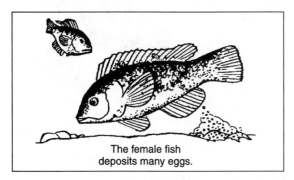

The female fish deposits many eggs.

Figure A

The male fish deposits a liquid over the eggs.

Figure B

1. What do we call the depositing of unfertilized eggs? _____

2. What name is given to the mass of fish eggs? _____

3. What is the liquid in Figure B called? _____

4. What does the liquid contain? _____

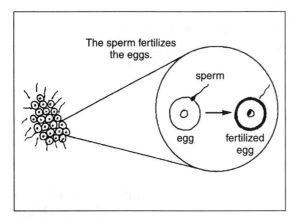

The sperm fertilizes the eggs.

sperm

egg   fertilized egg

Figure C

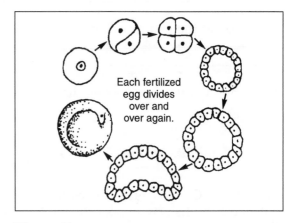

Each fertilized egg divides over and over again.

Figure D

5. How do the sperm reach the eggs? _____

_____

6. How many sperm cells fertilize one egg? _____

7. What is a fertilized egg called? _____

8. After a zygote divides many times what does it become? _____

9. What is a very young fish called? _____

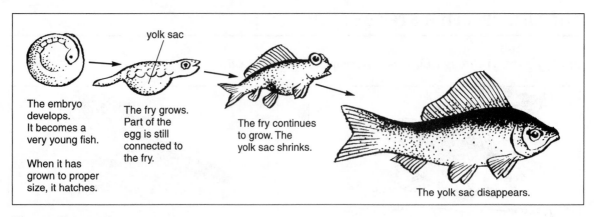

The embryo develops. It becomes a very young fish.

When it has grown to proper size, it hatches.

yolk sac

The fry grows. Part of the egg is still connected to the fry.

The fry continues to grow. The yolk sac shrinks.

The yolk sac disappears.

Figure E

10. What do we call the part of the egg that is connected to the fry? _____

_____

11. What is the job of the yolk sac? _____

_____

12. What happens to the yolk sac as the fry grows? _____

_____

13. Look at the end drawing of Figure E. What has happened to the fry? _____

_____

14. How will the fish now get its food? _____

SOME UNUSUAL BREEDING HABITS

The female <u>Convoy Fish</u> carries her fertilized eggs in her mouth until they hatch. The hatched baby fish swim near the mother. When there is danger, the babies dash back into the mother's mouth.

Figure F   *Convoy Fish*

Figure G *Sea Horse*

Figure H *Brook Stickleback Fish*

The Sea Horse is not a horse at all. It just looks like a horse. The male Sea Horse takes care of fertilized eggs. He carries them in a pouch on his belly.

Only birds build nests? Not so! The Brook Stickleback Fish builds a nest from plant strands. Then the female lays her eggs in the nest.

Pacific Salmon are born in freshwater streams, but they spend most of their lives in the salt water of the Pacific Ocean. However, they return to freshwater to spawn. Adult salmon swim upstream, some as many as 2,000 miles (3,200 kilometers). They battle rushing currents and leap over waterfalls as high as 10 feet (3 meters). They spawn only once and die soon afterward.

Figure I

1. Why are male sea horses unusual?

   _____

   _____

   _____

2. How do baby Convoy Fish avoid

   danger? _____

   _____

3. How many times does a Pacific Salmon spawn? _____

4. How are Brook Stickleback Fish like birds? _____

5. Are Pacific Salmon adapted to both freshwater and salt water? _____

6. How do you know? _____

   _____

FILL IN THE BLANK

Complete each statement using a term or terms from the list below. Write your answers in the spaces provided.

outside   yolk sac   spawning
fry   externally   sperm
roe   fish   feeds
milt

1. In most fish, fertilization and embryo development take place _____ .

2. In external fertilization, the egg is fertilized _____ the female's body.

3. A mass of fish eggs is called _____ .

4. The depositing of unfertilized fish eggs is called _____ .

5. The liquid deposited over the eggs by a male fish is called _____ .

6. Milt contains millions of _____ .

7. A very young fish is called a _____ .

8. The part of the egg connected to a hatched fry is called the _____ .

9. The yolk sac _____ the developing fry.

10. When a yolk sac is used up, the fry has become a bigger _____ .

MATCHING

Match each term in Column A with its description in Column B. Write the correct letter in the space provided.

	Column A		Column B
_____	1. external fertilization	a)	the depositing of fish eggs
_____	2. roe	b)	mass of fish eggs
_____	3. spawning	c)	method for most fishes
_____	4. milt	d)	developing fish
_____	5. fry	e)	liquid containing fish sperm
_____	6. yolk	f)	external
_____	7. outside	g)	food material

Use the clues to complete the crossword puzzle. Highlight the words you used in the lesson.

CLUES

ACROSS

1. Young fish
4. A part of your foot
5. Group of animals that live in water
7. The kind of fertilization found in humans
10. "RIGHT _____ !"
11. Not down
12. The part of an egg that contains food
13. Female sex cell
16. A developing plant or animal
17. Mass of sperm
20. Male sex cell
22. Moving air
23. Abbreviation for southeast

DOWN

1. The joining of a male and female gamete
2. 365 days
3. Kind of pickle
4. Prefix meaning "three"
6. Not sad
8. Not even one
9. "_____ apple a day . . ."
12. This toy has its "ups and downs"
14. Female
15. Fish eggs
16. Take in food
18. Same as 11 Across
19. Abbreviation for doctors
20. Abbreviation for street
21. Self

TRUE OR FALSE

In the space provided, write "true" if the sentence is true. Write "false" if the sentence is false.

_____   **1.** All fish reproduce the same way.

_____   **2.** Most fish eggs are fertilized outside the female's body.

_____   **3.** Most fish embryos develop inside the female's body.

_____   **4.** Male fish deposit eggs.

_____   **5.** Male fish produce roe.

_____   **6.** Roe are fish eggs.

_____   **7.** Milt contains millions of sperm.

_____   **8.** One sperm fertilizes one egg.

_____   **9.** A young fish is called a fry.

_____   **10.** A young fish is called a fry until it uses up its yolk sac.

WORD SCRAMBLE

Below are several scrambled words you have used in this Lesson. Unscramble the words and write your answers in the spaces provided.

1. TILM   _____

2. TYZOEG   _____

3. KOYL   _____

4. WGNPSAIN   _____

5. ALMSNO   _____

REACHING OUT

A fish is a vertebrate [VUR-tuh-brayt]. A vertebrate is an animal with a backbone.

Do you have a backbone? _____

Are you a vertebrate? _____

Figure J

116

How do frogs reproduce and develop?

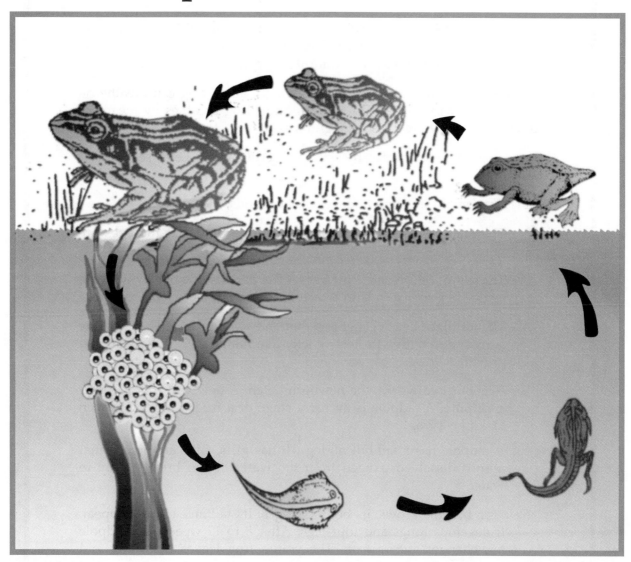

KEY TERMS

amphibian: animal that lives part of its life in water and part on land

tadpole: early stage of a frog

gills: organs that absorb dissolved oxygen from water

metamorphosis: changes during the stages of development of an organism

LESSON 18 | How do frogs reproduce and develop?

A frog is a vertebrate. It is called an **amphibian** [am-FIB-ee-un]. A frog, like all amphibians, spends part of its life in water and part on land. Its early life is spent in water. Its adult life is spent mostly on land.

A young frog can breathe only in water. An adult frog can breathe on land, as well as in water. On land an adult frog breathes by means of lungs. In water, it breathes through its skin.

Frogs reproduce in water by external fertilization. Let's see how it happens:

1. The female lays hundreds, even thousands, of eggs.

2. The male releases his sperm at the same time that the eggs are released.

3. The sperm cells fertilize the eggs. (The parents do not stay with the eggs. They return to land.) Each fertilized egg is called a zygote.

4. The fertilized eggs cluster together and swell greatly. They become surrounded with a protective jelly-like material. Cell division and growth follow.

5. In about two weeks, the newborn hatch. A newborn frog is called a **tadpole**. A tadpole is an early stage of a frog. It does not even look like a frog.

    A tadpole has a tail but no legs. It has **gills**. Gills are organs that absorb dissolved oxygen from the water. It can breathe only in water.

6. As a tadpole grows, its body changes. Its tail and gills disappear. It develops lungs and four legs. After 8 to 12 weeks, the tadpole has changed into a frog. The young frog leaves the water and becomes a land and water animal.

From tadpole to frog is a complete change in body form. A change in body form that certain animals go through as they grow is called **metamorphosis** [met-uh-MOWR-fuh-sis].

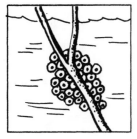

Figure A   *A fertilized egg mass*

A mass of fertilized eggs looks like jelly with white and black spots. The black spots are zygotes. The white spots are yolks.

1. What will the yolks be used for? _____

2. Frog fertilization is _____ .
   _{internal, external}

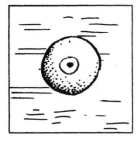

Figure B   *A fertilized egg*

3. What is a fertilized egg called? _____

4. What fertilizes a frog egg? _____

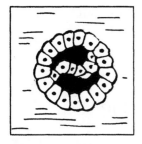

Figure C   *Embryo*

5. The zygote becomes an embryo. For this to happen, the zygote _____ over and over again.

Figure D   *Newly hatched offspring*

6. What do we call a young frog offspring? _____

7. A tadpole can live only _____ .
   in water, on land

8. A tadpole breathes by means of _____ .
   lungs, gills

Figure E *Full-sized tadpole*

Figure F *Hind legs form*

Figure G *Front leg "buds" start to appear*

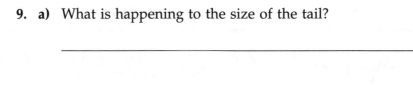

9. a) What is happening to the size of the tail?

_____

b) Of what use is the tail?

_____

10. Compare Figures D, E, and F. In what ways is the tadpole

changing? _____

_____

11. What is developing inside that will take the place of the

gills? _____

12. Compare the hind legs in Figure F to those in Figure G. What

is happening to the hind legs? _____

13. What is starting to happen to the size of the tail?

_____

14. Look at Figure H. Is any part of the tail left? _____

15. How many legs does the adult frog have? _____

16. Does the frog still have gills? _____

17. a) What does the adult frog use for breathing on land?

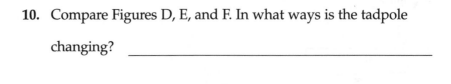

b) What does the adult frog use for breathing in water?

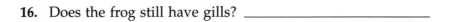

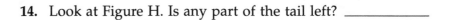

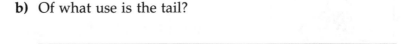

Figure H *Adult frog*

Frogs are important because they eat insects.

A frog has a very long sticky tongue. It is rolled up in its mouth, but it springs out suddenly to catch insects.

Figure I

How large is a frog? Frog size varies greatly.

Figure J

This young Western Tree Frog is less than 2½ cm (one inch) long.

Even when full grown, it will measure less than 5 cm (2 inches) long.

Figure K

An adult bullfrog is about 18–20 cm (7–8 inches) long.

121

FILL IN THE BLANK

Complete each statement using a term or terms from the list below. Write your answers in the spaces provided. Some words may be used more than once.

external   tail   amphibian
lungs   legs   tadpole
does not   vertebrates   metamorphosis
water   land and water   gills

1. Animals with backbones are called _____ .

2. A frog is a special kind of vertebrate called an _____ .

3. An amphibian spends its early life in _____ and its adult life on _____ .

4. Frogs reproduce in _____ .

5. Frogs reproduce by _____ fertilization.

6. A young frog offspring is called a _____ .

7. A tadpole _____ look like a frog.

8. A tadpole has a _____ and _____ .

9. A tadpole does not have _____ or _____ .

10. A change in body form that happens to some animals as they grow is called

_____ .

Study the stages in the life cycle of a frog in Figure L. Indicate the correct order of development by writing letters A to F in the spaces provided.

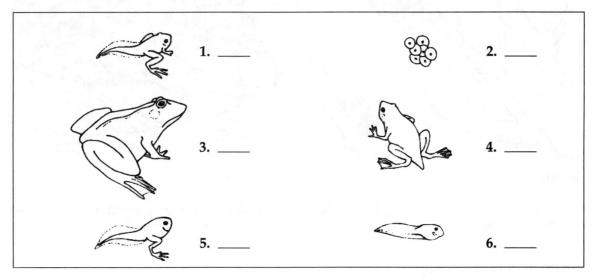

1. ____   2. ____

3. ____   4. ____

5. ____   6. ____

Figure L

122

MATCHING

Match each term in Column A with its description in Column B. Write the correct letter in the space provided.

	Column A		Column B
_____	**1.** frog	**a)**	young frog offspring
_____	**2.** tadpole	**b)**	for breathing on land
_____	**3.** gills	**c)**	complete change in body form
_____	**4.** lungs	**d)**	lives on land and in water
_____	**5.** metamorphosis	**e)**	for the breathing in water

WORD SEARCH

The list on the left contains words that you have used in this Lesson. Find and circle each word where it appears in the box. The spellings may go in any direction; up, down, left, right, or diagonally.

METAMORPHOSIS

TADPOLE

ZYGOTE

GILLS

FROG

LUNGS

EGG

VERTEBRATE

MILT

TESTES

OVARY

M	I	G	Y	R	A	G	S	T	A	R	F	M
L	E	T	E	M	A	G	G	Y	Z	O	T	I
Z	Y	T	G	G	N	I	O	S	I	G	T	L
Y	E	L	A	U	T	L	E	S	O	M	A	P
G	A	I	L	M	E	L	R	P	H	E	D	O
O	M	M	I	G	O	S	O	M	A	T	P	E
T	S	E	L	J	L	R	Y	R	A	V	O	L
E	T	L	V	S	G	R	P	E	Z	G	L	M
D	T	A	P	O	G	V	A	H	Y	A	E	T
E	T	E	S	T	E	S	Y	G	O	R	F	E
R	A	V	E	B	L	E	S	O	T	S	E	L
V	E	R	T	E	B	R	A	T	E	V	I	N
T	L	E	O	P	S	G	Y	L	L	I	M	S

In the space provided, write "true" if the sentence is true. Write "false" if the sentence is false.

_____ **1.** Frogs have backbones.

_____ **2.** Frog eggs are fertilized inside the female's body.

_____ **3.** Frogs reproduce on land.

_____ **4.** A female frog protects her fertilized eggs.

_____ **5.** A young frog offspring is called a fry.

_____ **6.** Tadpoles breathe only in water.

_____ **7.** A tadpole has gills and a tail.

_____ **8.** A frog has gills and a tail.

_____ **9.** A frog can breathe on land and in water.

_____ **10.** A frog can breathe through its skin.

REACHING OUT

Figure M

A frog's hind legs are very powerful. They are also webbed.

1. What <u>two</u> important things does a frog do with his hind legs?

    _____

2. What does the webbing help a frog do?

    _____

3. Do any people sometimes wear things that look like frog feet?

    _____

4. When and why are they worn?

    _____

    _____

What is meiosis?

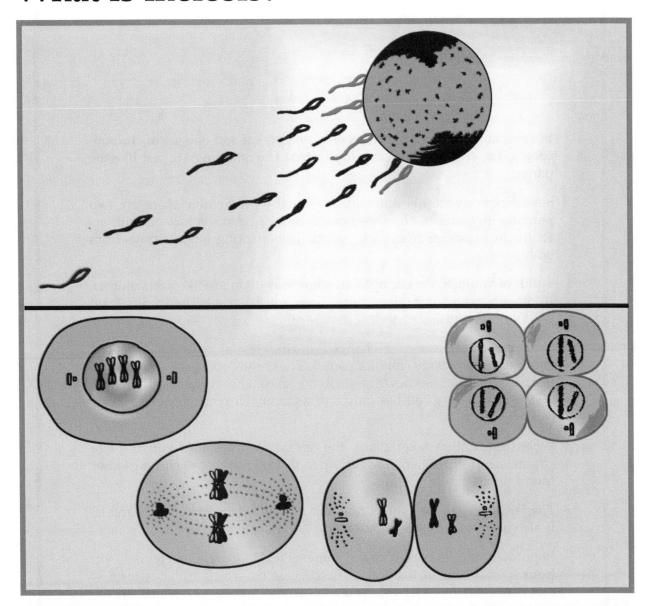

KEY TERMS

gametes: reproductive cells

meiosis: process by which gametes form

LESSON 19 | What is meiosis?

In asexual reproduction there is only <u>one</u> parent and <u>one</u> set of chromosomes. The chromosomes are duplicated. The offspring are just like the parent.

Sexual reproduction is different. In sexual reproduction, there are <u>two</u> parents—<u>two</u> sets of chromosomes. A new organism is formed with one set of chromosomes from each parent. The offspring inherits traits from both parents.

Think of yourself, for example. In some ways you are like your mother. In other ways you are like your father. You have inherited traits from both your parents.

How are chromosomes exchanged during sexual reproduction? The chromosomes of body cells are paired. The chromosomes of sex cells are not paired. Chromosomes of sex cells are single chromosomes. Therefore, a sperm or an egg cell has only half as many chromosomes as a body cell.

When fertilization takes place, the sperm chromosomes join the egg chromosomes. Together, they add up to the full number of chromosomes found in body cells.

The fertilized egg, or zygote, now has chromosomes from both parents. It also has traits from both parents.

Reproductive cells also are called **gametes** [GAM-eets]. Gametes develop from special cells in the body. The process by which gametes form is called **meiosis** [my-OH-sis]. You can see the process of meiosis on the next page.

1. Original cell with two pair of chromosomes. One member of each chromosome pair is from one parent. The other chromosome is from the other parent.

2. First, the amount of DNA doubles. Then the members of every pair of similar chromosomes pair with one another. Frequently, paired chromosomes will exchange segments in a process called crossing over.

3. Spindle fibers form in the cell. The pairs of similar chromosomes attach to the spindle fibers.

4. The members of the similar pairs of chromosomes separate.

5. The cell splits in two, with each cell containing one double chromosome from each pair.

6. Spindle fibers form in each new cell. Double chromosomes attach to the spindle fibers.

7. The double chromosomes split. One chromosome goes to each side of the cell. Then, each cell splits.

8. Four cells are produced. Each cell has one chromosome from each pair.

Figure A

Body cells are produced by mitosis. But sperm and egg cells do not form this way. Reproductive cells are formed by meiosis. Each gamete has only <u>half</u> the usual number of chromosomes. But when the sperm and egg join, the zygote has the full number of chromosomes.

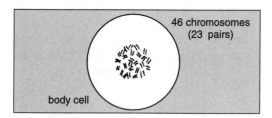

Figure B

A human body cell has 46 chromosomes. The chromosomes are paired. So there are 23 pairs.

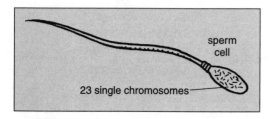

Figure C

Each human sperm cell has 23 single chromosomes.

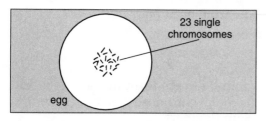

Figure D

Each human egg cell has 23 single chromosomes.

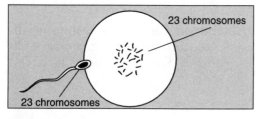

Figure E

Fertilization links the gamete chromosomes.

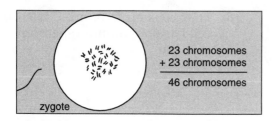

Figure F

The zygote, then, has a total of 46 chromosomes. 23 are from the mother, 23 are from the father.

The zygote starts to divide after fertilization. It divides by mitosis. It divides over and over again as it develops.

TRUE OR FALSE

In the space provided, write "true" if the sentence is true. Write "false" if the sentence is false.

_____   **1.** The chromosomes of body cells are paired.

_____   **2.** The process by which gametes form is meiosis.

_____   **3.** A human body cell has 23 chromosomes.

_____   **4.** In sexual reproduction, an offspring inherits traits from only one parent.

_____   **5.** Fertilization links the chromosomes of gametes.

_____   **6.** Every organism has the same number of chromosomes.

_____   **7.** Spindle fibers form twice during meiosis.

_____   **8.** A gamete has the same number of chromosomes as a body cell.

_____   **9.** A gamete has twice the number of chromosomes as a body cell.

_____   **10.** A frog gamete has 13 chromosomes. Every frog body cell, then, has 26 chromosomes.

FILL IN THE BLANK

Complete each statement using a term or terms from the list below. Write your answers in the spaces provided.

one set	parent	just like
half	traits	paired
two	meiosis	

1. In asexual reproduction there is only one _____ .

2. In asexual reproduction, _____ of chromosomes is passed on from parent to offspring.

3. In asexual reproduction, offspring are _____ the parent.

4. In sexual reproduction, there are _____ parents. Offspring inherit

   _____ from both parents.

5. Gamete cells are produced by cell division called _____ .

6. A sperm or egg cell has only _____ as many chromosomes as a body cell.

7. Chromosomes in a body cell are _____ .

Scientists often study fruit flies because they have large chromosomes whose genes are easy to see.

- Every body cell of a fruit fly has 8 chromosomes.
- Every fruit fly gamete (sperm or egg) has 4 chromosomes.

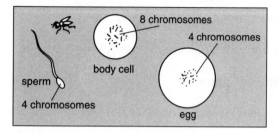

Figure G

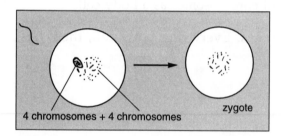

Figure H   *A sperm fertilizes an egg.*

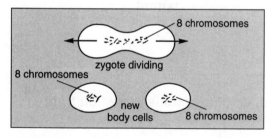

Figure I   *The zygote divides. Then each new cell divides.*

1. Body cells reproduce by a process called _____ .

2. Gamete cells are formed by a process called _____ .

3. **a)** How many chromosomes does the egg cell of a fruit fly have? _____

   b) Sperm cells? _____

4. What do chromosomes control?

   _____

5. How many chromosomes does a fruit fly zygote have? _____

6. How many chromosomes will each body cell have? _____

7. The offspring will have traits of the mother and the father. Why? _____

   _____

   _____

REACHING OUT

Why must a gamete have only one half the number of chromosomes found in body cells?

_____

_____

How are living things classified?

KEY TERMS

taxonomy: science of classifying living things

kingdom: largest classification group

phylum: classification group made up of related classes

genus: classification group made up of related species

species: group of organisms that look alike and can reproduce among themselves

LESSON 20 | How are living things classified?

How many different kinds of plants and animals can you name? Twenty? Thirty? . . . Fifty? It may be hard to believe, but we share this planet with millions of different kinds of organisms. So far, about 1½ million different kinds of organisms have been identified. In addition, about 6,000 more are being discovered each year. Some scientists believe the number to be more than 10,000,000.

How do we keep track of so many different kinds of organisms? Biologists classify living things into groups. Members of the same group are alike in certain important ways. The science of classifying living things is called **taxonomy** [tak-SAHN-uh-mee].

Carolus Linnaeus [lun-NAY-us] developed a classification system in 1735. Linnaeus is credited as the founder of modern taxonomy. He grouped organisms according to what they looked like. Organisms that looked alike were grouped together. Today, we consider other things too, like cell organization, chemical make-up, ancestors, and the way an organism develops before it is born.

CLASSIFICATION GROUPS   Today, living things are classified into seven major classification groups. Organisms that are classified in the same group are alike in some ways. The more alike organisms are, the more groups they share.

The largest classification group is the **kingdom**. A kingdom contains the largest number of different organisms. Members of a kingdom share only a few traits or characteristics. In fact, members of the same kingdom may not look alike at all. Take the flea and the elephant, for example. Do they look alike? Of course not! Yet, they belong to the same kingdom.

Each kingdom is then divided into smaller and smaller groups. They are the **phylum** [FY-lum], the class, the order, the family, the **genus** [JEE-nus], and the **species** [SPEE-sheez]. As a group becomes smaller, the members become more similar.

Think of the classification system as an upside-down pyramid. The kingdom is the largest part. It has the most room, so it can hold the greatest number of organisms—all plants, or all animals, or all protists.

As you move down the pyramid, each room gets smaller. It can hold fewer and fewer members; however, the members have more traits in common. They begin to be more alike.

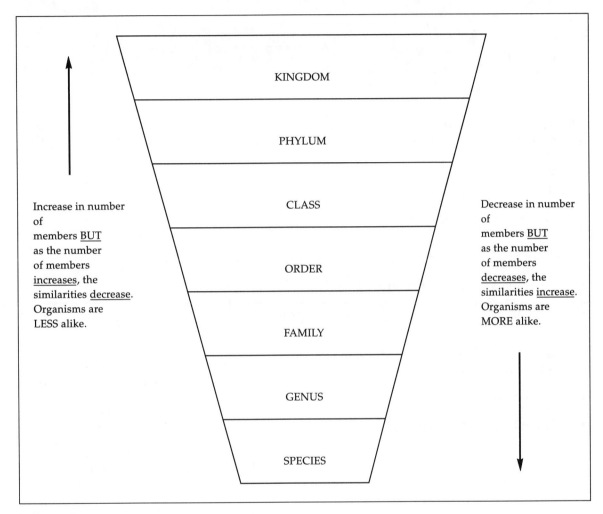

KINGDOM

PHYLUM

CLASS

ORDER

FAMILY

GENUS

SPECIES

Increase in number of members <u>BUT</u> as the number of members <u>increases</u>, the similarities <u>decrease</u>. Organisms are LESS alike.

Decrease in number of members <u>BUT</u> as the number of members <u>decreases</u>, the similarities <u>increase</u>. Organisms are MORE alike.

Figure A

The species has the smallest space in the classification pyramid. It is only large enough for one kind of organism—only humans, or only elm trees, or only robins.

Members of a particular species are very similar. Organisms in the same species look alike and can reproduce among themselves . . . Are there any differences? Certainly! But they are mostly individual differences—like the differences between two people, or two elm trees, or two robins.

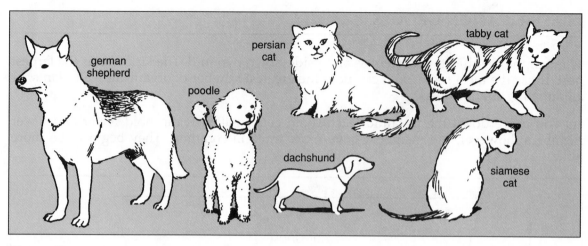

Figure B

Some organisms, like dogs and cats, are classified into an even smaller group—the breed. But regardless of the breed, all dogs belong to one species. All cats belong to another species.

TRACING SIMILARITIES

A member of a particular classification group has traits that are the same as

- all members of its own group, as well as

- all members of the broader classification groups.

For example: A member of a particular phylum has traits that are similar to

- all members of that phylum, as well as

- all members of its kingdom.

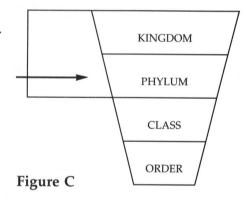

Figure C

Use Figures A and C to help you answer the following questions.

CLASSIFICATION GROUP	**SHARES SOME CHARACTERISTICS WITH THESE GROUPS**
1. a. a member of a <u>class</u>	_____
2. b. a member of an <u>order</u>	_____
3. c. a member of a <u>species</u>	_____
	_____
	_____

2. Which are more similar:

    a. members of an order or members of a family? _____

    b. members of an order or members of a phylum? _____

3. Which has more members:

    a. a phylum or a family? _____

    b. a genus or a family? _____

4. Which group has fewer members:

    a. a family or genus? _____

    b. an order or a phylum? _____

5. a. Which group has the most members? _____

    b. Which group has the fewest members? _____

Study the table showing the classification of four organisms. Answer the question in the spaces provided.

Table 1   Classification of Organisms

	Dandelion	**Dog**	**Wolf**	**Human**
Kingdom	Plantae	Animalia	Animalia	Animalia
Phylum	Tracheophyta	Chordata	Chordata	Chordata
Class	Angiospermae	Mammalia	Mammalia	Mammalia
Order	Asterales	Carnivora	Carnivora	Primates
Family	Compositae	Canidae	Canidae	Hominidae
Genus	*Taraxacum*	*Canis*	*Canis*	*Homo*
Species	*officinale*	*familiaris*	*lupus*	*sapiens*

6. How many groups do wolves and humans share? _____

7. How many groups do wolves and dogs share? _____

8. a. Which two organisms are the most similar? _____

    b. How do you know? _____

9. a. Which organism is least like the other three? _____

    b. How do you know? _____

10. In what kingdom is the dandelion classified? _____

NAMING ORGANISMS

Carolus Linnaeus developed a system of classification. He also developed a system for naming organisms that is still used today. His system is called <u>binomial nomenclature</u> (bi-NOH-mee-uhl NOH-muhn-klay-chur). In this system, each kind of organism is given a two-part scientific name. The first part is the name of the genus in which the organism is classified. The second part is the name of the species. For example, the scientific name for a dog is *Canis familiaris*. (See Table 1). When a scientific name is written, the genus name is italicized or underlined.

Answer the following questions about scientific names. You may need to study Table 1 to answer some of the questions.

1. What are the two parts of a scientific name? _____

2. a. What is the system for naming organisms that is used today called? _____

    _____

    b. Who developed this system? _____

3. a. What is the scientific name for humans? _____

    b. What is the scientific name for wolves? _____

    c. What is the scientific name for dandelions? _____

4. Why do scientists use scientific names? _____

    _____

5. What is wrong with the way this scientific name is written? *canis Familiaris?*

    _____

At one time all organisms were classified in either the plant or animal kingdom. Today most scientists accept the five kingdom classification system.

Here is a description of the five kingdoms.

Moneran [muh-NER-un] Kingdom

Single-celled organisms. Unlike members of the other four kingdoms, monerans do not have a nucleus. Bacteria are examples of monerans.

Protist Kingdom

Contains many different kinds of organisms. Most protists are single-celled. Some are simple, many-celled organisms. Some are like plants. Others are like animals. Algae are one example of protists.

Fungi Kingdom

Do you like mushrooms? Mushrooms are one kind of fungi. Most fungi are made up of many cells. Some have only one cell. Fungi absorb food from their environment.

Plant Kingdom

Plants have many cells. Plant cells have a cell wall and contain chlorophyll. Plants use chlorophyll to make their own food.

Animal Kingdom

You are probably most familiar with members of the animal kingdom. Animals are made up of many cells. They take in food from the outside.

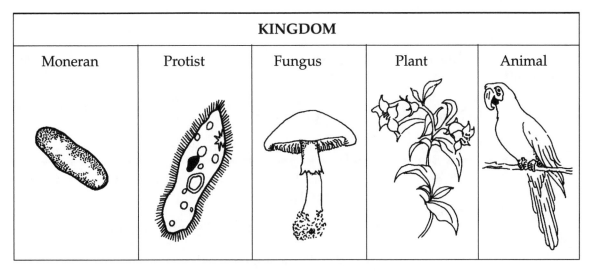

KINGDOM				
Moneran	Protist	Fungus	Plant	Animal

Figure D

Complete each statement using a term or terms from the list below. Write your answers in the spaces provided. Some words may be used more than once.

single-celled   bacteria   many
nucleus   chlorophyll fungi
five   absorb   outside

1. Most scientists accept the _____ kingdom classification system.

2. Plants have _____ cells.

3. Fungi _____ food from their environment.

4. _____ are examples of monerans.

5. Animals take in food from the _____ .

6. Mushrooms are one kind of _____ .

7. Most protists are _____ .

8. Plants use _____ to make their own food.

9. Animals have _____ cells.

10. Monerans do not have a _____ .

MATCHING

Match each term in Column A with its description in Column B. Write the correct letter in the space provided.

Column A	Column B
_____ 1. Monerans	a) one example of protists
_____ 2. mushroom	b) all have many cells and take in food from the outside
_____ 3. animals	c) all have many cells and make their own food
_____ 4. algae	d) all are single-celled
_____ 5. plants	e) one kind of fungi

What are monerans?

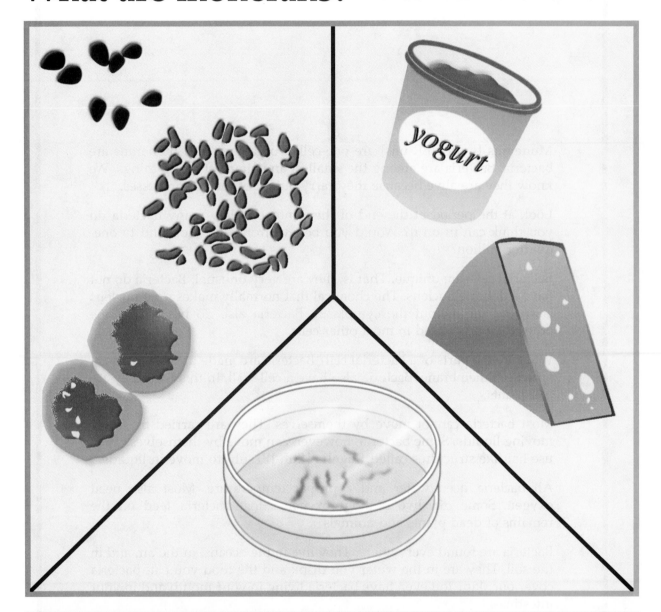

KEY TERMS

monerans: single-celled organisms that do not have a nucleus

flagella: hairlike structures that bacteria use to move

coccus: round bacteria

bacillus: rod-shaped bacteria

spirillum: spiral-shaped bacteria

LESSON 21 | What are monerans?

Monerans [muh-NER-uns] are one-celled organisms. All monerans are bacteria. Bacteria are among the smallest and simplest living things. We know they are alive because they carry out all of the life processes.

Look at the period at the end of this sentence. How many bacteria do you think can fit on it? Would you believe from fifty thousand to one-quarter <u>million</u>?

Bacterial cells are unique. That is, they are <u>very</u> unusual. Bacteria do not have a definite nucleus. The chemical that normally makes up a nucleus is spread throughout the cytoplasm. Bacteria also do not have some other cell parts found in most other cells.

What are the parts of a bacterial cell? Bacteria are made up of cytoplasm, and a cell membrane. Bacteria also have a cell wall. In this way, they are like plants.

Most bacteria cannot move by themselves. They are carried by air or moving liquids. Some bacteria, however, can move by themselves. They use hairlike structures called **flagella** [fluh-JEL-uh] to move in liquids.

All bacteria need water and a proper temperature. Most also need oxygen. Some can live without oxygen. Most bacteria feed on the remains of dead plants and animals.

Bacteria are found everywhere. They live in the oceans, in the air, and in the soil. They are in the water you drink and the food you eat. Bacteria cover our skin. You even have bacteria living in your mouth and in your intestines!

KINDS OF BACTERIA

Bacteria [sing. *bacterium* (bak-TIR-ee-um)] are found in three basic shapes. They may be round, rod-shaped, or spiral. A round bacteria is called **coccus** [sing. KAHK-us (pl. *cocci* KAHK-sy)]. A rod-shaped bacterium is called a **baccillus** [sing. buh-SIL-us (pl. *bacilli* buh-SIL-y)]. A spiral shaped bacterium is called a **spirillum** [sing. spy-RIL-um (pl. *spirilla* spy-RIL-uh)]. Some cocci form pairs and chains. Others grow in grapelike bunches. Some bacilli also form pairs or chains. But they do not form bunches. Spirilla only live as single cells.

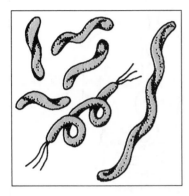 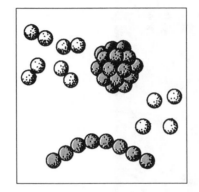 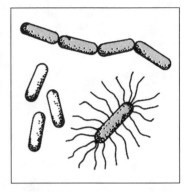

Figure A   **Figure B**   **Figure C**

Look at Figures A, B, and C above. Use letters to answer the following questions.

1. Which illustration shows   **a)** cocci?   _____

                                      b) spirilla?   _____

                                      c) bacilli?   _____

2. One of the bacteria shown can move by itself.

    a) Which one? _____

    b) Draw a picture of this bacteria in the space below.

    ```



    ```

3. **a)** Which illustration shows bacteria that live <u>only</u> as single cells? _____

    b) Which illustration shows bacteria growing in grapelike clusters?

    _____

    c) What kinds of bacteria may form pairs and chains? _____

Bacteria are important for many reasons.

Figure D *Bacteria are used to make certain foods. Some are used to make cheese, yogurt, vinegar, and sauerkraut.*

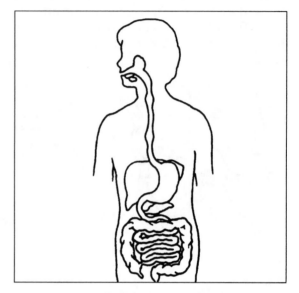

Figure E *You have bacteria living in your digestive tract! These bacteria help you digest food. Some also produce nutrients such as vitamin K.*

Figure F *Some bacteria are nature's recyclers! They break down the remains of dead organisms. When a plant or animal dies, bacteria of decay feed upon the organism and break down their food into simple substances such as carbon. These substances are reused as nutrients by other living things.*

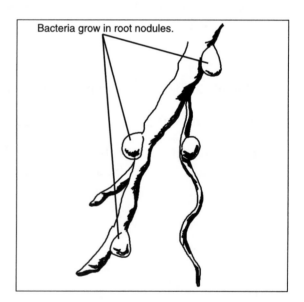

Bacteria grow in root nodules.

Figure G *Plants cannot use nitrogen gas directly. Bacteria in soil and in the root nodules of some plants change nitrogen in the air into a form plants can use. Animals get the nitrogen they need when they eat plants.*

You have just learned that bacteria can be helpful. Of course, they can also be harmful. Some bacteria cause diseases in plants and animals. Bacteria also cause food to spoil.

Figure H *Spoiled food*

Bacteria cause food to spoil. Foods that are easily spoiled by bacteria include fish, milk, fruits, and vegetables.

Figure I *Diseased plant*

Some bacteria cause tumorlike growths in plants. Each year millions of dollars are lost from crop damage caused by bacterial diseases.

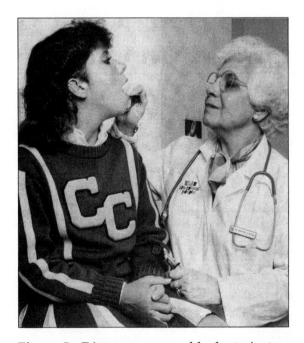

Figure J *Diseases are caused by bacteria, too.*

Bacteria cause many diseases in people. Some of the diseases caused by bacteria are tuberculosis, strep throat, cholera, pneumonia, food poisoning, tetanus, diptheria, and meningitis.

143

HELPFUL AND HARMFUL BACTERIA

Complete the chart by giving one way in which bacteria are helpful and harmful for each example.

Helpful and Harmful Bacteria

	EXAMPLE	HELPFUL	HARMFUL
1.	Humans		
2.	Plants		
3.	Foods		

TRUE OR FALSE

In the space provided, write "true" if the sentence is true. Write "false" if the sentence is false.

_____ **1.** Foods that are easily spoiled by bacteria include fish and milk.

_____ **2.** Some spirilla form chains.

_____ **3.** Some bacteria can live without oxygen.

_____ **4.** Most bacteria cannot move by themselves.

_____ **5.** All bacteria are harmful.

_____ **6.** Bacterial cells have a cell wall.

_____ **7.** Plants can use nitrogen gas directly.

_____ **8.** Cocci are rod-shaped bacteria.

_____ **9.** Bacteria have many cells.

_____ **10.** Some bacteria live in your digestive tract.

GROWING BACTERIA

A single bacterium and groups of bacteria are <u>microscopic</u>. You need a microscope to see them. However, under proper growth conditions, bacteria form <u>colonies</u>. Large colonies of bacteria can be seen with the unaided eye. You can grow bacteria in a special broth called agar. Agar is a food source for bacteria.

What You Need (Materials)

- sterile agar in 3 sterile covered **petri** [PEE-tree] dishes

- grease pencil

How To Do The Experiment (Procedure)

1. Number each petri dish (1 to 3) on the bottom. Keep a record of what you do to each petri dish.

2. Petri dish #1: Open the cover quickly and sprinkle some dust or fine soil on the dish. Quickly close the cover.

   Petri dish #2: Quickly open the petri dish. Wait a few seconds, then replace the cover.

   Petri dish #3: Do nothing. Most importantly, do not remove the cover.

3. Place all 3 petri dishes in a warm dark place. After 2 days, observe each petri dish.

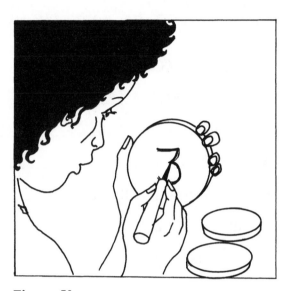

Figure K

What You Learned (Observations)

1. Which petri dishes showed signs of bacterial growth? _____

   _____

2. Which petri dish showed no sign of bacterial growth? _____

Something To Think About (Conclusions)

1. Why was there no growth in petri dish #3? _____

   _____

2. Why do bacteria grow in agar? _____

   _____

FILL IN THE BLANK

Complete each statement using a term or terms from the list below. Write your answers in the spaces provided.

spiral	monera	rod-shaped
breaking down dead organisms	diseases	water and a proper temperature
pairs and chains	nucleus	bacilli
cocci	round	life processes
spirilla	flagella	

1. Some cocci form _____ .

2. Some bacteria use _____ to move.

3. Bacteria carry out all the _____ .

4. Bacteria do not have a true _____ .

5. Bacteria belong to the kingdom _____ .

6. All bacteria need _____ .

7. The three shapes of bacteria are _____ , _____ , and

   _____ .

8. Round bacteria are called _____ . Rod-shaped bacteria are called

   _____ . Spiral bacteria are called _____ .

9. Bacteria can cause many human, plant, and animal _____ .

10. One way that bacteria can be helpful is by _____ .

REACHING OUT

Some unusual bacteria called blue-green bacteria are able to make their own food. Blue-green bacteria contain the green substance chlorophyll. Chlorophyll is needed in order for blue-green bacteria to make their own food. How are blue-green bacteria like plants?

_____

_____

_____

_____

_____

What are protists?

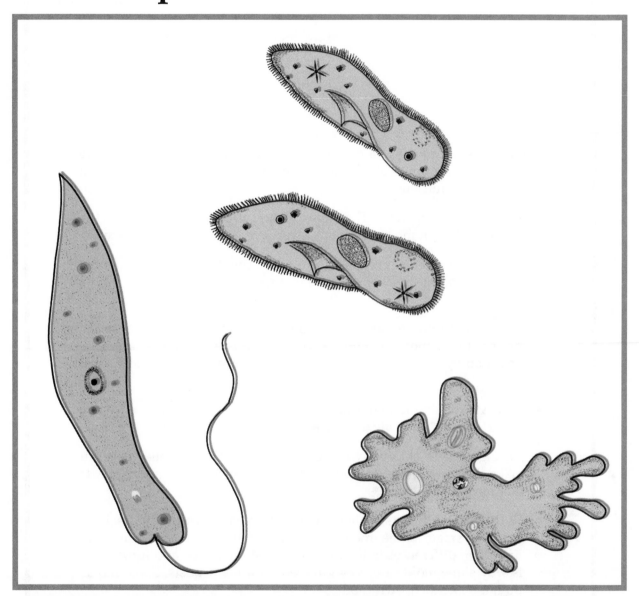

KEY TERMS

protist: simple organisms that have a true nucleus

protozoan: one celled animal-like protist that cannot make its own food

algae: plant-like group of protists that can make their own food

slime mold: protists that have two life stages

LESSON 22 | What are protists?

The protists are a kingdom of very simple organisms. Most **protists** are made up of only one cell. Some are many-celled. Unlike bacteria, all protists have a nucleus.

Most protists live in water. They live in lakes, streams, ponds, and in the ocean. Some live in moist soil. Some even live in the bodies of animals.

There are three large groups of protists. The three groups are the **protozoans** [proht-uh-ZOH-uns], the **algae** [AL-jee], and the **slime molds**.

- **PROTOZOANS** are one-celled animal-like organisms. Like animals, protozoans cannot make their own food. Most protozoans can move about on their own in search of food. Protozoans do not have a cell wall.

- **ALGAE** are simple plant-like organisms. Most algae have only one cell. However, some algae have many cells and grow to be very large.

  All algae have chlorophyll [KLAWR-uh-fil]. Like plants, they can make their own food. However, algae do not have the special parts that most plants have.

- **SLIME MOLDS** are simple organisms that have two life stages. During one stage, slime molds are similar to fungi. But there are important differences in the life cycles of slime molds and fungi. That is why slime molds are now grouped with the protists. Slime molds do not make their own food.

PROTOZOANS

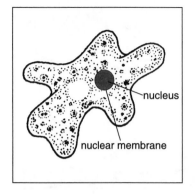 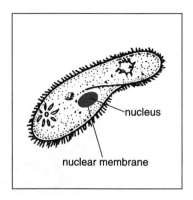 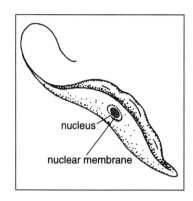

Figure A *Amoeba*   **Figure B** *Paramecium*   **Figure C** *Trypanosoma*

Three common protozoans are shown above. Use the illustrations and what you have read about protists to answer the questions below.

1. Protozoans are _____ celled organisms.
   <u>single, many</u>

2. Do protozoans make their own food? _____
   <u>yes, no</u>

3. Do most protozoans move by themselves? _____
   <u>yes, no</u>

4. Do protozoans have a cell wall? _____
   <u>yes, no</u>

5. Look carefully at each nucleus. The nucleus of a protozoan

   a) has a membrane around it. _____
   <u>yes, no</u>

   b) is spread out in the cytoplasm. _____
   <u>yes, no</u>

ALGAE

All algae have chlorophyll. However, not all algae are green. Algae also have other substances, which give them different colors. Some algae are green. Other algae are red, brown, or golden-brown. Algae are found in many watery places . . .

- You may have seen algae as a green scum on a still pond or lake.

- The green "fuzz" on some moist rocks is algae. Do not walk on algae covered rocks. They are very slippery.

- Have you ever seen green water in a swimming pool? The green in the water is algae. Some swimming pools have a real "algae problem."

- Most algae are found in the ocean. Have you ever seen seaweed washed up on a beach? Seaweed is another name for brown algae.

How important are algae? As important as life itself . . . Most of the oxygen we breathe is produced by one-celled algae that float near the surface of water.

Figure D

Answer the questions below.

1. What substance do all algae have that enables them to make their food? _____

2. What are four colors of algae? _____

3. Why are one-celled algae important? _____

SOME INTERESTING FACTS ABOUT SLIME MOLDS

Have you ever seen the movie, *The Blob?* During part of its life cycle, a slime mold looks like a slimy, shapeless "blob."

A slime mold does not make its own food. It "oozes" over the forest floor in search of food. A slime mold absorbs its food.

Slime molds are very unusual protists. At one time slime molds look like fungi. But during another stage, slime molds look and act like amoebas. You have just learned that amoebas are protozoans. A slime mold may be made up of many amoeba-like cells. Or, sometimes slime molds are just a mass of cytoplasm with thousands of nuclei.

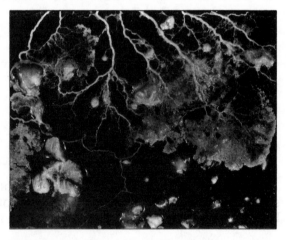

Figure E

4. How does a slime mold get its food? _____

5. What organisms do slime molds resemble during different times of their life cycles?

_____

You are looking through a microscope . . . Well, not really! But <u>imagine</u> that you are.

Figure F on page 152 shows what you are "seeing"—15 different kinds of protists.

Anton van Leeuwenhoek [An-tun van LAY-vun-hook] was a pioneer microscope-maker. He lived more than 300 years ago. Van Leeuwenhoek was the first person to see tiny protists. Guess what he called them? "Wee beasties!" Now they have fancy scientific names.

Can you identify these protists from their descriptions? Sure you can! It's easier than you think. BUT, there is a C-A-T-C-H. Pronouncing some of the names won't be easy. Almost everyone has trouble . . . even your teacher. So don't be discouraged. In fact, it should be lots of FUN!

READY . . . SET . . . GO!

Match each protist shown in Figure F to its description in the chart. Write the correct letter in the space in the chart. (Hint: You may wish to fill in and check off the ones you are sure of first.)

	Scientific Name	Looks Like	Letter
1.	diatom	a triangle	
2.	heteronema	a snail	
3.	protochrysis	a kidney bean	
4.	chlamydononas	a drop with whiskers	
5.	paramecium	a footprint	
6.	vorticella	a flower with a telephone cord for a stem	
7.	loxodes	a cucumber with a bite taken out	
8.	radiolarian	a lacy snowflake	
9.	lacrymaria	a swan	
10.	anarma	a purse with hairy corners	
11.	heliozoan	the sun	
12.	spyrogyra	a tube with coiled worms inside	
13.	forminiferan	a bunch of hairy grapes	
14.	glenodinium monensis	a hamburger	
15.	trachelocera conifer	a baseball bat	

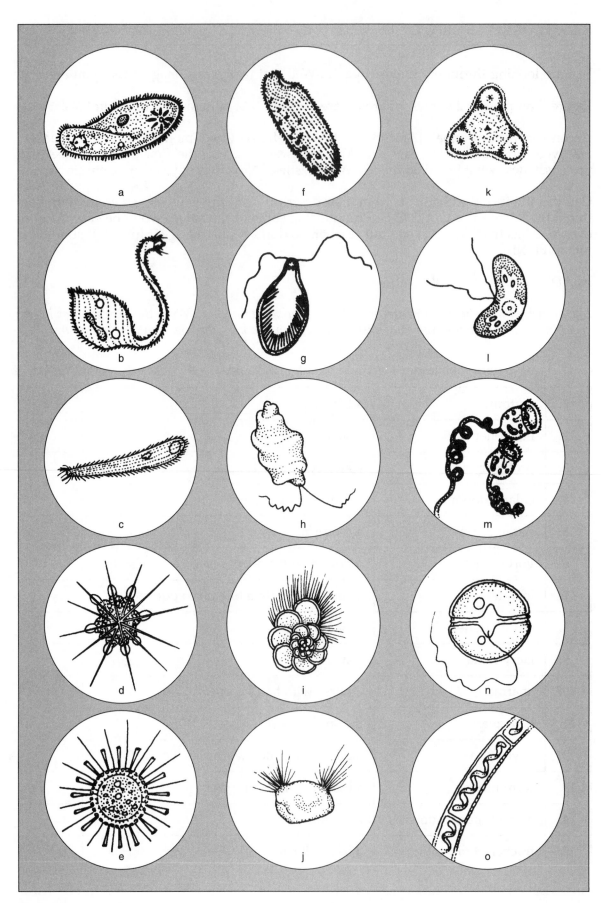

Figure F *The World of Protists*

Lesson **23**

What are fungi?

fungi: plantlike organisms that lack chlorophyll

chitin: hard substance that makes up the cell walls of fungi

stalk: stemlike part of a mushroom

cap: umbrella-shaped part of a mushroom

gills: structures that produce mushroom spores

rhizoids: rootlike structures that anchor fungi

LESSON 23 | What are fungi?

Do you think that the mushrooms in a salad are related to the lettuce? Many people think that mushrooms are plants. However, mushrooms are not plants. Mushrooms are one kind of **fungi**.

Fungi are like plants in some ways. For example, fungi often grow well in soil like most plants. The cells of fungi have cell walls. Plants have cell walls too. And like some plants, fungi reproduce by spores.

Scientists, however, have discovered that fungi and plants are not really very much alike. Let's see how they are different:

- The cells of fungi do not have chloroplasts or chlorophyll. Therefore, fungi cannot make their own food as plants do. Fungi must get food from their surroundings. Most fungi feed upon dead organisms.

- The cell walls of fungi are made up of a hard substance called **chitin** [KYT-in]. You may remember that the cell wall of a plant is made up of cellulose.

- Fungi often have large cells with many nuclei.

- Fungi grow well in moist, dark places.

You now know that mushrooms are a kind of fungus. The fungi kingdom also includes yeasts and molds.

Yeasts

If you have ever baked bread, you probably used yeast to make the bread rise. Yeasts are single-celled, colorless fungi. They grow well where sugar is present. Yeasts use sugar for food.

Figure A *Yeasts produce carbon dioxide as bread dough is baked. It causes bubbles to form. As the bubbles form, the bread rises.*

Molds

Molds are common kinds of fungi. They grow on bread, fruit, and even leather. Most molds look like a mass of threads. Some threads are rootlike structures called **rhizoids** [RY-zoidz]. They hold the mold to the bread. Nutrients move up through the rhizoids to other parts of the mold. Other threads produce spores. (See Lesson 13.)

Figure B *Black bread mold is a common mold.*

Mushrooms

Like molds, mushrooms are made up of many threads. However, the threads are packed closely together. You probably can recognize a mushroom by its shape. The stemlike part of a mushroom is called the **stalk**. At the top is an umbrella-shaped **cap**. The underside of the cap is lined with **gills**. The gills produce spores. The rootlike structures at the base of the stalk are **rhizoids.**

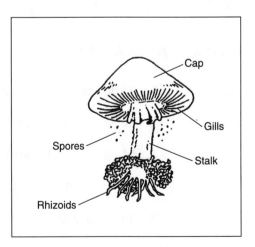

Figure C *Parts of a mushroom.*

155

COMPLETE THE CHART

Answer the questions by putting a "YES" or "NO" in the space provided.

		Plants	Fungi
1.	Are all many-celled?		
2.	Do most grow well in sunny places?		
3.	Do most have cell walls made up of chitin?		
4.	Do their cells have chlorophyll?		
5.	Are their cells surrounded by a cell wall?		
6.	Are their cell walls made up of chitin?		
7.	Can they make their own food?		
8.	Do they often have large cells with many nuclei?		
9.	Do they grow by producing threadlike structures?		
10.	Do their cells have chloroplasts?		
11.	Do most feed on dead organisms?		
12.	Do they all produce spores?		
13.	Do they grow well in dark, moist places?		
14.	Are most microscopic?		
15.	Are their cell walls made up of cellulose?		

TRUE OR FALSE

In the space provided, write "true" if the sentence is true. Write "false" if the sentence is false.

_____ **1.** Molds often grow on bread and fruits.

_____ **2.** Yeasts grow well where sugar is present.

_____ **3.** Mushrooms do not have cell walls.

_____ **4.** The gills of a mushroom produce spores.

_____ **5.** Yeasts are made up of many threads.

PARTS OF A MUSHROOM

Label the diagram of a mushroom. Use the labels: **cap, gills, rhizoids, spores,** and **stalk.**

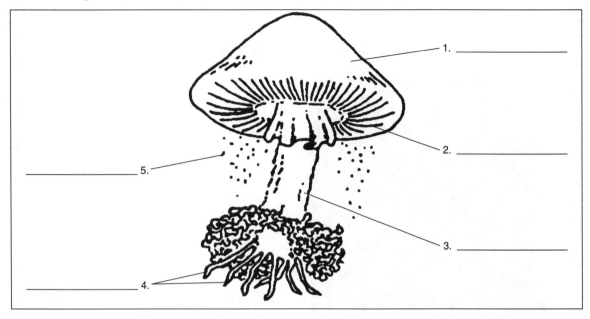

1. _____

2. _____

3. _____

4. _____

5. _____

Figure D

REACHING OUT

Many people enjoy eating mushrooms. However, some mushrooms are poisonous. Why should you <u>NEVER</u> pick and eat wild mushrooms?

_____

_____

_____

_____

_____

Figure E

SCIENCE *EXTRA*

Tracking Animals

To protect animals from harm or even extinction, scientists have to understand how the animals live. What do they eat? Where do they sleep? How far do they travel in a day or in a month?

Wild animals are difficult to study. They fly, run, swim, and burrow where scientists cannot see them. Scientists can use technology to find out where the animals go and what they do.

First, scientists capture an animal. Then, they attach a radio transmitter to the animal. The transmitter runs on batteries. It may be tied to a collar on a mountain lion. It may be glued to the shell of a sea turtle. The scientists then release the animal into its natural surroundings.

Radio receivers pick up signals from the transmitters. Scientists track an animal by listening to its signal. They also use computers and recorders to track the signals. In this way, scientists can keep track of animals. They do not have to be near them. Sometimes, scientists track animals using satellites. The satellites pick up the radio signals. Then, they send them to a receiver.

Scientists can also put other devices on animals. Some devices measure an animal's blood pressure and temperature. Others measure the weather around an animal. The radio transmitters send this information to a receiving station.

If you could track animals that live in your area, which animals would you study? What would you try to find out? With the information you would gather, you would be able to do what scientists do. You might find out what is causing problems for the animals. Then you might work on ways to protect them from harm.

Plants

Lesson

What are the characteristics of plants?

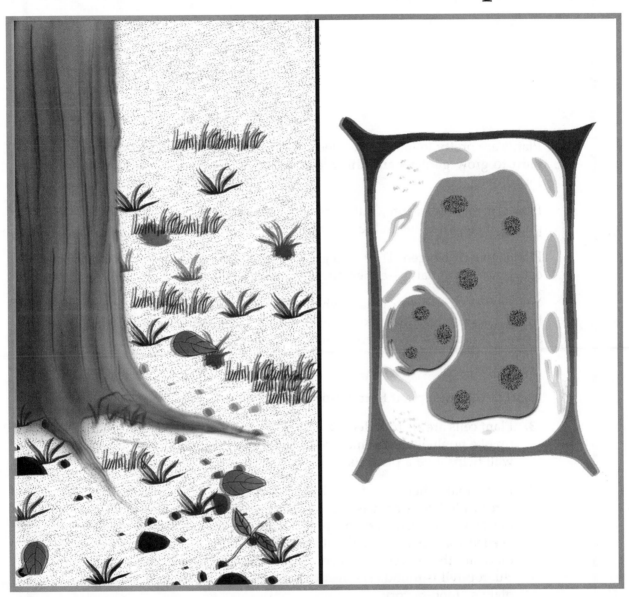

KEY TERMS

cellulose: nonliving substance that makes up the plant cell walls

chloroplast: structure in the cells of a green plant that store chlorophyll

chlorophyll: green pigment needed by a plant in order to make its own food

seed: reproductive structure

LESSON 24 | What are the characteristics of plants?

Plants are pretty to look at. We decorate our homes with them. We work hard to grow gardens and thick lawns. Sometimes, we give flowers to loved ones.

Plants give us pleasure, but they are even more important. Without plants, there could be no life on Earth!

Plants give us oxygen. They also give us food. In fact, all of the food we eat comes from plants—either <u>directly</u> or <u>indirectly</u>.

You probably recognize a plant when you see one. However, as you have learned, some algae look like plants. How can you tell if an organism is a plant? All plants have these characteristics:

1. Plants have many cells.

2. The cells of a plant form <u>tissues</u> and <u>organs</u>.

3. Plant cells are surrounded by a rigid cell wall. This cell wall is made up of a nonliving substance called **cellulose** [SEL-yoo-lohs]. The cell wall helps give a plant its stiffness.

4. Plants make their own food. The cells of green plants contain structures called **chloroplasts** [KLAWR-uh-plasts]. Chloroplasts contain the green substance, **chlorophyll** [KLAWR-uh-fil]. Chlorophyll is needed for a plant to make its own food. Most food-making takes place in the leaves of green plants. This is where most of the chlorophyll is found. The leaf of the plant can be thought of as the plant's "food-factory."

PLANT PHYLA

The plant kingdom is divided into two large groups called phyla—the <u>vascular</u> [VAS-kyuh-lur] plants and the <u>nonvascular</u> plants.

VASCULAR PLANTS belong to the division *Tracheophyta* [TRAY-kee-uh-fyt-uh]. Vascular plants have stems, roots, and leaves. All of the plants in this division also have a vascular system. A vascular system is a system of connecting tubes. These tubes carry water and dissolved nutrients to all parts of the plant.

The vascular plants are the most complex plants. Most of the plants you know belong to this group of plants. The vascular plants include ferns, trees, roses, and other flowering plants.

NONVASCULAR PLANTS belong to the division *Bryophyta* [BRY-uh-fyt-uh]. The plants in this division are very simple. Nonvascular plants do not have true roots, stems, or leaves. They also have no vascular system.

Examples of nonvascular plants include the mosses, liverworts, and hornworts. You have probably seen moss. It grows in moist, shaded areas. It is commonly found on the soil under trees, in sidewalk cracks, and on walls.

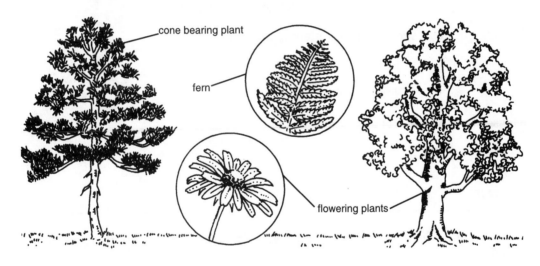

Figure A

Figure A shows some common plants. Study the diagrams. Then answer the questions.

1. a) The plants shown are _____ plants.

vascular, nonvascular

    b) They _____ have a tube system to transport water and

do, do not
        dissolved nutrients.

2. These plants belong to the phylum _____ .

Bryophyta, Tracheophyta

3. Do tracheophytes have true roots, stems, and leaves? _____

4. Tracheophytes are _____ plants.

simple, complex

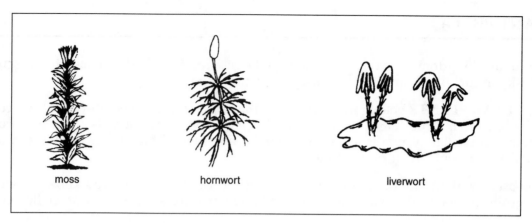

Figure B

moss   hornwort   liverwort

5. Name the plants shown in Figure B. _____ _____ _____

6. They are _____ plants.
   _{vascular, nonvascular}

7. These plants belong to the division _____ .
   _{Bryophyta, Tracheophyta}

8. Do bryophytes have true roots, stems, and leaves? _____
   _{yes, no}

9. Bryophytes are _____ plants.
   _{simple, complex}

10. Mosses and liverworts are examples of _____ .
    _{bryophytes, tracheophytes}

MATCHING

Match each term in Column A with its description in Column B. Write the correct letter in the space provided.

	Column A		**Column B**
_____	1. chlorophyll	**a)**	made up of cellulose
_____	2. vascular plants	**b)**	grow in moist, shaded areas
_____	3. plant kingdom	**c)**	have stems, roots, and leaves
_____	4. mosses	**d)**	green substance
_____	5. cell wall	**e)**	divided into two large groups

COMPLETE THE CHART

Some of the characteristics listed in the chart below are true of tracheophytes only. Some are true only of bryophytes. Some of the characteristics are found in tracheophytes and bryophytes. Answer the questions by putting a "YES" or "NO" in the space provided.

		DIVISION	
		Tracheophyta	Bryophyta
1.	Are they vascular plants?		
2.	Do their cells have cell walls?		
3.	Are they nonvascular plants?		
4.	Do they have a system of connecting tubes?		
5.	Do their cells contain chloroplasts?		
6.	Are they made up of many cells?		
7.	Do they have true roots, stems, and leaves?		
8.	Can they make their own food?		
9.	Do mosses belong to this group?		
10.	Do trees belong to this group?		
11.	Are they very simple plants?		
12.	Are they complex plants?		
13.	Do liverworts belong to this group?		
14.	Do roses belong to this group?		
15.	Do ferns belong to this group?		

In Lesson 13 you learned that some plants reproduce by spores. For example, ferns are common spore plants.

Most plants, however, reproduce by **seeds**. A seed is a reproductive structure. All seed plants are vascular plants.

Biologists classify seed plants into two groups. One group has uncovered seeds. This group is called gymnosperms [JIM-nuh-spurms]. The other group's seeds have tough outer coverings. This group is called angiosperms [AN-jee-uh-spurms].

The most common and best known of the gymnosperms are the evergreens. Evergreens produce cones. Their seeds are in the cones. Evergreens also have special leaves called needles. The needles stay green throughout the year. Pines, cedars, and spruces are gymnosperms.

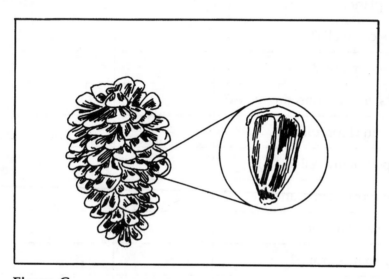

Figure C

Angiosperms are the flowering plants. Most of the common plants you see everyday are angiosperms. In some angiosperms, such as roses and tulips, the flowers are very noticeable. In others, the flowers are very small. Grasses, oak trees, and corn are angiosperms. Have you ever seen their flowers?

GYMNOSPERM OR ANGIOSPERM?

Classify each of the plants below as a gymnosperm or an angiosperm. Then write whether each has an uncovered seed or a covered seed.

Figure D   *Corn plant*

1. A corn plant is a _____ .
   <u>gymnosperm, angiosperm</u>

   It has _____ seeds.
   <u>uncovered, covered</u>

Figure E   *Pine tree*

2. A pine tree is a _____ .
   <u>gymnosperm, angiosperm</u>

   It has _____ seeds.
   <u>uncovered, covered</u>

Figure F   *Spruce tree*

3. A spruce tree is a _____ .
   <u>gymnosperm, angiosperm</u>

   It has _____ seeds.
   <u>uncovered, covered</u>

Figure G   *Tulip plant*

4. A tulip is a _____ .
   <u>gymnosperm, angiosperm</u>

   It has _____ seeds.
   <u>uncovered, covered</u>

5. The most common gymnosperms are the _____ .
   <u>evergreens, grasses</u>

6. Most plants reproduce by _____ .
   <u>spores, seeds</u>

7. Angiosperms have _____ .
   <u>cones, flowers</u>

8. Evergreens have special leaves called _____ .
   <u>needles, flowers</u>

Complete each statement using a term or terms from the list below. Write your answers in the spaces provided. Some words may be used more than once.

dissolved nutrients	phyla	do not
bryophytes	leaves	food
roots	complex	stems
oxygen	cellulose	vascular
cell wall	water	chloroplasts
many		

1. There can be no life without plants. Plants give us _____ and

   _____ .

2. The plant kingdom is divided into two large groups called _____ .

3. Plants make their own _____ .

4. Plant cells are surrounded by a rigid _____ .

5. Plant cells contain structures called _____, which contain chlorophyll.

6. A _____ system is a system of connecting tubes.

7. In some plants, _____ and _____ are carried throughout the plant by a vascular system.

8. Vascular plants are more _____ than nonvascular plants.

9. Vascular plants have true _____, _____, and

   _____ .

10. Nonvascular plants _____ have true roots, stems, or leaves.

11. Most food-making takes place in the _____ of green plants.

12. Plants have _____ cells.

13. Mosses, liverworts, and hornworts are _____ .

14. Chlorophyll is needed for a plant to make its own _____ .

15. The cell wall is made up of _____ .

What are roots, stems, and leaves?

KEY TERMS

tissues: groups of cells that look alike and do the same job

root hairs: tiny hairlike structures that help a root absorb more water

root cap: covers and protects the tip of a root

simple leaf: leaf that has all its leaf blades in one piece

compound leaf: leaf that has its leaf blade divided into smaller leaflets

stomata: tiny openings in a leaf

LESSON 25 | What are roots, stems, and leaves?

All plants are made up of many cells. The cells of vascular plants, or tracheophytes, form **tissues**. Tissues are groups of cells that look alike and do the same job. The tissues of vascular plants also are organized into organs. Roots, stems, and leaves are three plant organs.

ROOTS

Most roots grow underground. Roots hold a plant in the soil. Roots also take in water and dissolved minerals from the soil, and in some plants, food is stored in the roots.

STEMS

The main job of stems is to support the leaves. Stems also are important organs for carrying materials between the roots and leaves. In some plants, stems store food too. For example, the stems of sugarcane store large amounts of sugar.

LEAVES

How would you describe a leaf? You would probably say that a leaf is green, flat, and thin. Leaves are green because they contain chlorophyll. Chlorophyll is needed for a plant to make its own food. Food-making takes place in the leaves of a plant.

PARTS OF A ROOT

- Roots are tube-like structures made up of three layers. Many tiny, hairlike structures called **root hairs** come out from the outer layer. Root hairs help a root to absorb more water.

- The middle layer stores water and food for the root.

- The inner layer is made up of transport, or vascular tissue.

The tip of a root is covered by a **root cap**. The root cap protects a root tip from damage as the tip grows into the soil.

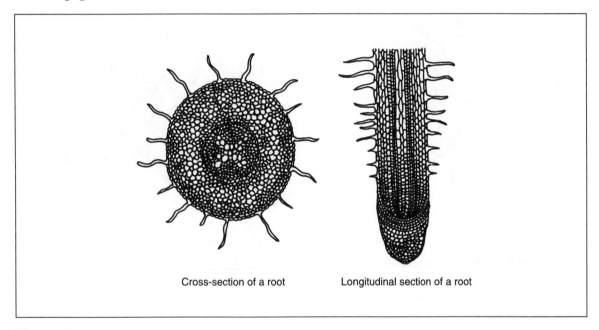

Cross-section of a root   Longitudinal section of a root

Figure A

Answer the questions below by studying the reading selection and Figure A.

1. How many layers make up a root? _____

2. On what layer are root hairs found? _____

3. What is the job of root hairs? _____

4. What is the inner layer made up of? _____

5. What structure covers a root tip? _____

There are two main kinds of root systems. They are <u>fibrous</u> [FY-brus] <u>root systems</u> and <u>taproot systems</u>. Fibrous roots are made up of many thin, branched roots. A taproot system has one large root. Many small, thin roots grow out from the large root.

*Identify the kind of root system shown in Figures B to E as a **fibrous root** or **taproot system**.*

Figure B   *Carrot*

6. _____

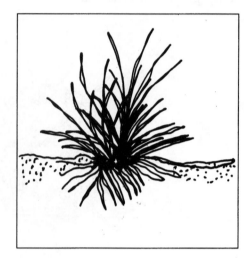

Figure C   *Grass*

7. _____

Figure D   *Wheat*

8. _____

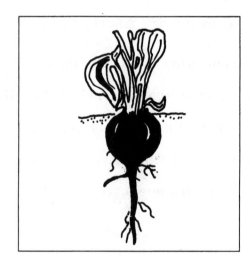

Figure E   *Radish*

9. _____

Read about each picture. Then answer the questions next to the picture.

There are two kinds of plant stems. They are <u>herbaceous</u> [hur-BAY-shus] <u>stems</u> and <u>woody stems</u>. Herbaceous stems are soft and green.

1. What are two kinds of plant stems? _____

   _____

2. What kind of stem does a tulip plant have?

   _____

3. Why do you think plants with herbaceous stems

   do not grow very tall? _____

   _____

Figure F

Woody stems are thick, hard, and rough. All trees have woody stems.

4. Woody stems _____ green.
   are, are not

5. What is the rough outer layer of a woody stem

   called? _____

Figure G

Both herbaceous stems and woody stems have transport tubes. One group of tubes carries materials up from the roots. The other carries material from the leaves downward throughout the plant.

6. What materials do transport tubes carry up from

   the roots? _____

7. What material do transport tubes carry down

   from the leaves? _____

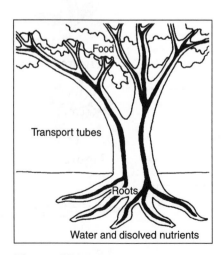

Food

Transport tubes

Roots

Water and disolved nutrients

Figure H

LEAVES

Most leaves have a stalk and a wide, flat part called the blade. The stalk attaches the leaf to the stem. Food-making takes place in the blade. Throughout the blade are many <u>veins</u>. The veins are made up of transport tissue.

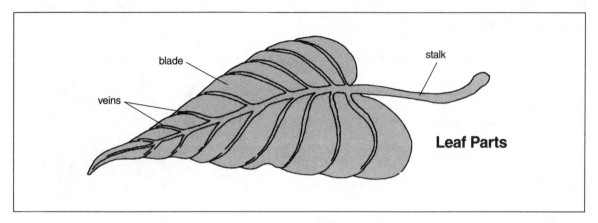

Figure I   *Leaf parts*

In some plants, the leaf blades are in one piece. This kind of leaf is a **simple leaf**. In other plants, the leaf is divided into smaller pieces called leaflets. This kind of leaf is a **compound leaf**.

Figure J   *Simple leaf*

Figure K   *Compound leaf*

Classify each leaf as simple (S) or compound (C).

_____   1.

Figure L

_____   2.

Figure M

_____   3.

Figure N

_____   4.

Figure O

_____   5.

Figure P

_____   6.

Figure Q

172

TRY THIS YOURSELF

Obtain a single leaf from a tree. Find the blade and stalk of your leaf. Check to see if your leaf has leaflets. Make a sketch of your leaf in the space below. Label each leaf part and write whether your leaf is a simple or compound leaf.

Figure R

GAS EXCHANGE

Stomates are tiny holes on the underside of a plant's leaves. Oxygen and carbon dioxide from the air enter a plant through the stomates. Along with water, oxygen and carbon dioxide also leave a plant through the stomates.

1. For what process do plants need oxygen?

   _____

2. What are the waste products of respiration?

   _____

3. What gas do plants need to make their own

   food? _____

Figure S

FILL IN THE BLANK

Complete each statement using a term or terms from the list below. Write your answers in the spaces provided. Some words may be used more than once.

compound	transport	woody
stomates	stems	food-making
protects	many	taproot

1. Herbaceous _____ are soft and green.

2. The veins of a leaf are made up of _____ tissue.

3. A carrot has a _____ system.

4. Fibrous roots are made up of _____ thin, branched roots.

5. Oxygen enters a plant through the _____ .

6. The main job of _____ is to support the leaves.

7. _____ takes place in the leaves.

8. The root cap _____ the root tip.

9. _____ stems are thick and rough.

10. A _____ leaf is divided into leaflets.

MATCHING

Match each term in Column A with its description in Column B. Write the correct letter in the space provided.

	Column A		Column B
_____	1. root hairs	**a)**	take in water and dissolved minerals
_____	2. fibrous root system	**b)**	carry materials throughout a plant
_____	3. simple leaf	**c)**	wide, flat part of a leaf
_____	4. roots	**d)**	covers the root tip
_____	5. transport tubes	**e)**	tiny holes
_____	6. stomates	**f)**	leaf blades are in one piece
_____	7. root cap	**g)**	root system of grass
_____	8. blade	**h)**	help a root absorb more water

How do living things get their energy?

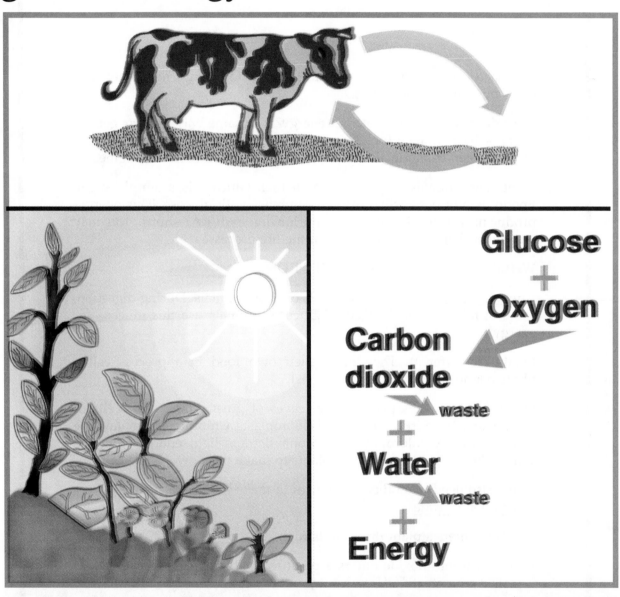

Glucose
+
Oxygen

Carbon
dioxide
→ waste
+
Water
→ waste
+
Energy

KEY TERMS

photosynthesis: food-making process in plants

respiration: process by which organisms release energy from food

LESSON 26 | How do living things get their energy?

You need energy to live, so do plants. <u>All</u> living things need energy to carry out their life processes.

How do plants and animals get energy? The same way our car gets its energy—by burning a fuel. Cars use gasoline as fuel. Energy is released when oxygen from the air combines with the gasoline in the engine.

Plants and animals use glucose as a fuel. Glucose is a simple sugar. Energy is produced when oxygen combines with glucose. This energy-producing process is called **respiration** [res-puh-RAY-shun]. You may remember that respiration is one of the life processes.

WHERE DOES GLUCOSE COME FROM?

Animals ingest food. They eat plants or other animals. During digestion, some of this food is changed to glucose. Animals use this glucose for energy.

Plants are different. They make their own food in a process called **photosynthesis** [foht-uh-SIN-thuh-sis].

Photosynthesis takes place in the leaves of green plants. The cells of green plants have structures called chloroplasts. Chloroplasts contain the green substance chlorophyll. Chlorophyll traps light energy from the sun. Light energy is needed for plants to make their own food.

Plants also need two other substances to make their own food — <u>water</u> and <u>carbon dioxide</u>.

- Water enters a plant through its roots.

- Most carbon dioxide enters a plant through tiny holes in its leaves called **stomata** [STOH-muh-tuh].

176

MORE ABOUT RESPIRATION AND PHOTOSYNTHESIS

The energy-producing process in living things is called respiration. Respiration is the release of energy by combining oxygen with digested food (glucose). Carbon dioxide and water are also produced. They are waste products of respiration.

A simple way to show respiration is this:

Glucose + Oxygen ⟶ **Carbon dioxide + water + energy**

waste   waste

The food-making process of green plants is called photosynthesis. Photosynthesis is also responsible for the production of oxygen.

Photosynthesis can be shown in this way:

Carbon dioxide + Water $\xrightarrow[\substack{\text{sunlight} \\ \text{(energy)}}]{\text{chlorophyll}}$ **Glucose + Oxygen**

Using the above information, answer the following questions.

1. What do we call the release of energy by living things? _____

2. Respiration is _____ in plants and animals.

different, the same

3. Two waste products produced by respiration are _____ and

_____ .

4. What fuel do living things use for energy? _____

5. What must link up with this fuel to produce energy? _____

6. Do plants take in food from the outside? _____

7. How do plants obtain food? _____

8. The food-making process of plants is called _____ .

9. Photosynthesis chemically combines two products. Name them.

   _____

   _____

10. Photosynthesis also requires energy and the green substance called

    _____ .

11. In which part of a plant does most photosynthesis take place?

    _____

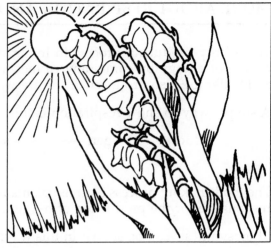

Figure A

Study Figure B carefully. Then answer these questions.

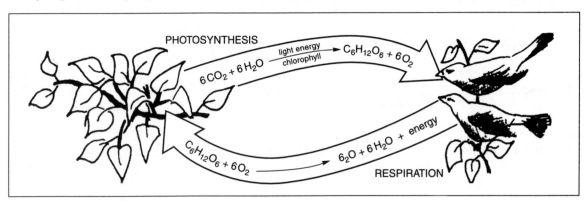

Figure B

12. In photosynthesis,

    a) what are the starting products? _____ and _____

    b) what are the end products? _____ and _____

13. In respiration,

    a) what are the starting products? _____ and _____

    b) what are the end products? _____ and _____

14. **a)** Is energy needed for photosynthesis to take place? _____

    b) Is energy produced by respiration? _____

15. Photosynthesis and respiration are _____ reactions.
    _{the same, opposite}

This picture shows photosynthesis taking place. When green plants are in sunlight, this is what happens:

water plus carbon dioxide **make** glucose and oxygen.

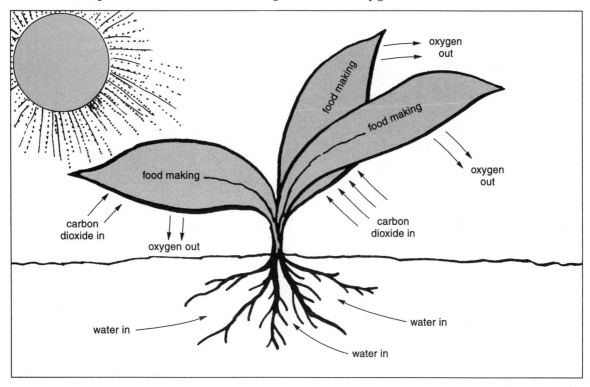

Figure C

Look at Figure C and then answer the following questions about photosynthesis.

1. Does photosynthesis take place in the leaves or in the roots? _____

2. What two materials must a plant take in for photosynthesis? _____

   _____

3. What else is needed for photosynthesis? _____

4. Where does the carbon dioxide gas come from? _____

5. What are the two things that photosynthesis makes? _____

6. What does a plant make as food? _____

7. Where does water enter a plant? _____

8. Where does carbon dioxide enter a plant? _____

9. Where does oxygen leave a plant? _____

10. What living things use the oxygen? _____

FILL IN THE BLANK

Complete each statement using a term or terms from the list below. Write your answers in the spaces provided. Some words may be used more than once.

chlorophyll	life processes	photosynthesis
glucose	carbon dioxide	oxygen
respiration	water	

1. The release of energy in living things is called _____ .

2. Respiration links up the simple sugar, _____ with the gas, _____ .

3. The waste products of respiration are _____ , _____ ,

   _____ .

4. Most of the energy released during respiration is used for the _____ .

5. The food-making process of plants is called _____ .

6. The products that chemically link up during photosynthesis are _____

   and _____ .

7. The substance in the leaves of green plants that is needed for photosynthesis is

   _____ .

REACHING OUT

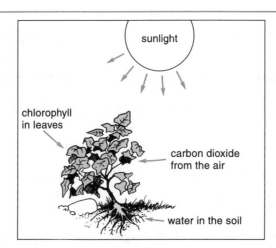

Figure D *Photosynthesis*

1. A plant makes its own food (glucose). Plants also make other nutrients. These nutrients are made of glucose, water, and minerals. How does a plant get minerals?

   _____

2. Algae have chlorophyll and carry out photosynthesis. How do algae get their food?

   (Hint: You can look back at Lesson 22 if you need help.)_____

   _____

180

What are the parts of a flower?

KEY TERMS

sepal: special leaves that support and protect the flower bud

stamen: male reproductive organ of a flower

pistil: female reproductive organ of a flower

anther: structure of a flower that produces pollen grains

imperfect: flower that has only pistils or stamens

pollen: male plant reproductive cell

ovule: small structures in which egg cells develop in a flower

perfect flower: flower that has both pistils and stamens

LESSON 27 | What are the parts of a flower?

Have you ever visited a botanical garden? A botanical garden is a place where you can see many different kinds of plants and many beautiful flowers. Not all plants have flowers. But in plants that do, the flower is the organ of sexual reproduction. A flower makes a plant's sex cells, or gametes. Male and female gametes join to produce seeds. Each seed may grow into a new plant.

Let us examine the parts of a flower. Check with Figure A on the next page as you read.

The bottom of a flower is surrounded by special leaves called **sepals** [SEE-puls]. Sepals support and protect the flower bud.

The petals of a flower are just inside the sepals. Petals are another kind of special leaf. Petals are the colorful leaves that protect the reproductive organs.

The most important parts of a flower are the **stamens** [STAY-muns] and the **pistil**.

STAMENS The stamen is the male reproductive organ of a plant. Most flowers have several stamens.

A single stamen has two parts: thread-like <u>filament</u> [FIL-uh-munt], and a knob-like **anther**. The anther is at the top of the filament.

An anther makes a powdery substance called **pollen**. Pollen grains are the male sex cells of a plant.

PISTIL The pistil is the female reproductive organ of a flower. It is located in the center of the flower. Some flowers have more than one pistil.

The lower part of a pistil bulges. This bulge is the <u>ovary</u> [OH-vur-ee]. An ovary contains one or more **ovules** [OH-vyoolz]. Each ovule has an egg cell. Egg cells are the female sex cells. Fertilization of an egg by a pollen grain takes place in the ovule. A fertilized egg becomes a seed. What will a seed produce?

Study Figure A. Then fill in the blanks.

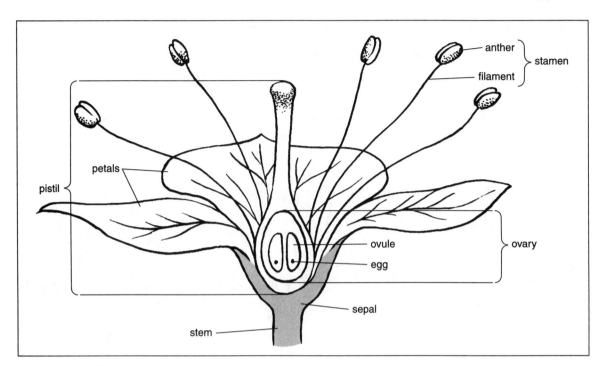

Figure A

1. Name the male reproductive organ of a plant. _____

2. Name the female reproductive organ of a plant. _____

3. Pollen is made by the _____ .

4. Eggs are made by the _____ .

5. How many stamens does this flower have? _____

6. **a)** Name the parts of a stamen. _____ _____

    b) Which part of a stamen makes the pollen? _____

7. What is the swollen part of a pistil called? _____

8. **a)** How many ovules does this ovary have? _____

    b) How many eggs are in each ovule? _____

9. What do the sepals do? _____

10. What do you call a flower's colorful special leaves? _____

PERFECT AND IMPERFECT FLOWERS

- Some flowers have only stamens. They are <u>male</u> or <u>staminate</u> [STAM-uh-nit] flowers.

- Some flowers have only pistils. They are <u>female</u> or <u>pistillate</u> [PIS-tuh-lit] flowers.

- A flower that has only pistils or stamens is an **imperfect** flower.

- Most plants have both stamens and a pistil. They are called **perfect** flowers.

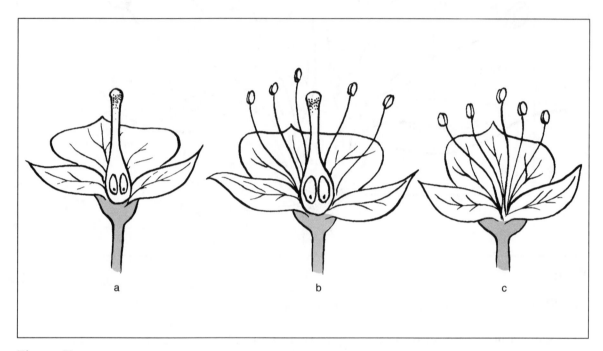

Figure B

Study the flowers in Figure B. Then answer the questions by placing the correct letter in the spaces provided.

1. Which is a staminate flower? _____

2. Which is a pistillate flower? _____

3. Which is a perfect flower? _____

4. Which flower makes only pollen? _____

5. Which flower makes only eggs? _____

6. Which flower makes both pollen and eggs? _____

7. Which flowers are imperfect flowers? _____

8. Look at the perfect flower. The stamens are _____ than the pistil.
   higher, lower

9. Which flower(s) has sepals? _____

10. Which flower(s) has/have petals? _____

FILL IN THE BLANK

Complete each statement using a term or terms from the list below. Write your answers in the spaces provided. Some words may be used more than once.

stamens	anther	egg
ovary	seed	ovules
pollen grains	pistil	gametes

1. A female sex cell is called an _____ .

2. Pollen grains are a male plant's _____ .

3. A male plant's sex cells are _____ .

4. The male part of a plant is made up of several _____ .

5. The part of a stamen that makes pollen is the _____ .

6. The entire female plant part is the _____ .

7. The swollen part of a pistil is the _____ .

8. An ovary has special parts called _____ .

9. An ovule contains a single _____ .

10. A fertilized plant egg cell becomes a _____ .

MATCHING

Match each term in Column A with its description in Column B. Write the correct letter in the space provided.

	Column A		Column B
_____	1. egg	a)	male plant sex cells
_____	2. pollen grains	b)	special leaves
_____	3. sepals	c)	female plant part where fertilization takes place
_____	4. ovule	d)	female sex cell
_____	5. anther	e)	have both stamens and a pistil
_____	6. imperfect flower	f)	makes the pollen
_____	7. flower	g)	reproductive organ of a plant
_____	8. perfect flowers	h)	has only a pistil or stamens

NAMING THE FLOWER PARTS

Label the parts of the flower in Figure C. Write the correct names on the spaces provided. Choose from the following words:

pistil	sepal	ovary
anther	ovule	stamen
petals	filament	egg

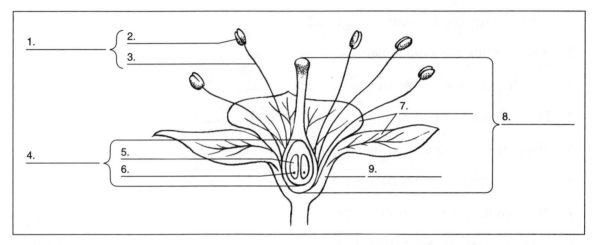

Figure C

TRUE OR FALSE

In the space provided, write "true" if the sentence is true. Write "false" if the sentence is false.

_____ **1.** A flower is an organ of sexual reproduction.

_____ **2.** All plants reproduce sexually.

_____ **3.** There are male plants and female plants.

_____ **4.** Some plants are both male and female.

_____ **5.** The most important parts of a flower are the stamens and the pistil.

_____ **6.** Stamens are female parts.

_____ **7.** Stamens make egg cells.

_____ **8.** An ovary has at least one ovule.

_____ **9.** One ovule has many eggs.

_____ **10.** A fertilized egg becomes a seed.

How does pollination occur?

KEY TERMS

pollination: movement of pollen from a stamen to a pistil

self-pollination: when pollen is carried from the stamen of one flower to the pistil of another flower on the <u>same</u> plant

cross-pollination: when pollen is carried from the stamen of a flower on one plant to the pistil of a flower on a <u>different</u> plant

LESSON 28 | How does pollination occur?

Before a pollen grain can fertilize an egg, pollen must move from a stamen to a pistil. The movement of pollen from a stamen to a pistil is called **pollination** [pahl-uh-NAY-shun].

There are two kinds of pollination—**self-pollination** and **cross-pollination.**

SELF POLLINATION happens in perfect flowers. A perfect flower has both stamens and a pistil. Self-pollination takes place when pollen from a stamen lands on the pistil of the same flower. Self-pollination also occurs when pollen from one flower is carried to the pistil of another flower on the same plant.

CROSS-POLLINATION is the transfer of pollen from a stamen of a flower on one plant to the pistil of a flower on another plant. Imperfect flowers pollinate only by cross-pollination. A male plant cannot pollinate itself. It has no pistil. A female plant makes no pollen. It must get pollen from some other plant. Perfect flowers also cross-pollinate. A perfect flower may give or receive pollen from another plant.

What carries pollen from plant to plant? There are two main carriers—wind and insects.

Wind   Pollen is like powder or dust. It is light in weight. Wind can carry pollen from the anther of one flower to the pistil of another flower of the same plant. Or, wind may carry pollen to the pistils of the same kind of plant far away. For example, wind helps pollinate corn plants.

Insects   Insects are attracted to flowers by their nice smell and bright colors. Insects also come to flowers to feed on a sweet liquid called nectar [NEC-tur].

Insects search inside flower petals for nectar. As they search, some pollen sticks to their bodies. The insects then may visit other flowers. There, the pollen may rub off the insects and onto other pistils.

Some insects that carry pollen are flies, gnats, butterflies, moths, and bees. Bees are the most important. In fact, bees are often "hired" to pollinate fruit trees! Hummingbirds and bats pollinate some flowers, too. And, sometimes, pollen is carried by water.

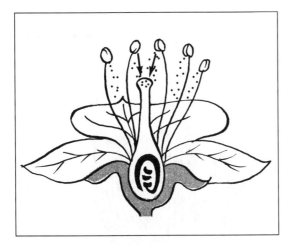

Figure A *Self-pollination*

Sometimes, gravity does the work in self-pollination. Pollen from stamens just drops onto the pistil of the same flower.

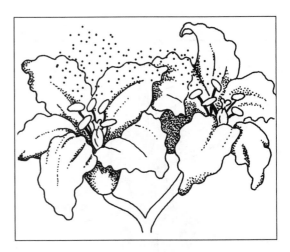

Figure B *Self-pollination*

Self-pollination also occurs when pollen from one flower is carried to the pistil of another flower on the same plant.

Figure C *Cross-pollination of imperfect flowers*

In cross-pollination, pollen from a flower of one plant is carried to a pistil of another plant.

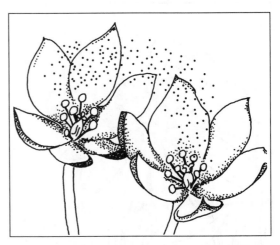

Figure D *Cross-pollination of perfect flowers*

Cross-pollination also occurs when pollen from one perfect flower is carried to the pistil of a perfect flower on another plant.

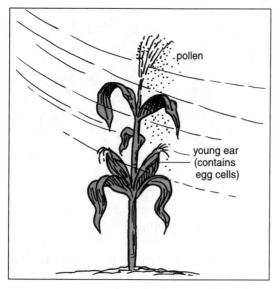

Figure E

Wind carries pollen from the stamens of a corn plant to the pistil. The female part is actually cornsilk.

Figure F

Moths and butterflies carry pollen from plant to plant—even cactus plants.

Figure G

Insects see colors that humans cannot. Some flowers have a colored "target" that is invisible to us.

Figure H

Insects do not see red. It looks black to them. Most red flowers are pollinated by birds.

UNDERSTANDING POLLINATION

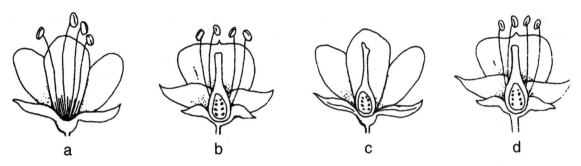

Figure I

Study the four flowers in Figure I. Then answer the questions below by placing the correct letter in the spaces provided.

1. Which flower makes pollen only? _____

2. Which flower make eggs only? _____

3. Which flowers can self-pollinate? _____

4. Which flower cannot pollinate any other flower? _____

5. Which flowers may flower **b** pollinate? _____

6. Which flower may flower **d** pollinate? _____

7. Which flower may flower **a** pollinate? _____

8. In which plants can fertilization take place? _____

Draw arrows in Figure I to show all of the ways these flowers can be pollinated.

MATCHING

Match each term in Column A with its description in Column B. Write the correct letter in the space provided.

Column A	Column B
_____ 1. nectar	a) pollinate most red flowers
_____ 2. birds	b) have only pistils or stamens
_____ 3. perfect flowers	c) important pollen carriers
_____ 4. imperfect flowers	d) sweet liquid
_____ 5. bees	e) have both stamens and pistils

FILL IN THE BLANK

Complete each statement using a term or terms from the list below. Write your answers in the spaces provided. Some words may be used more than once.

self-pollination	nectar	insects
wind	smell	stamen
pistil	cross-pollination	pollination
colors		

1. Corn plants are usually helped to pollinate by _____ .

2. The male part of a plant is the _____ .

3. The female part of a plant is the _____ .

4. Insects come to flowers to feed on _____ .

5. The transfer of pollen from any stamen to any pistil is called _____ .

6. The two kinds of pollination are _____ and _____ .

7. The pollination of a pistil by pollen from the same flower is called

    _____ .

8. The pollination of a pistil of one plant by pollen from a different plant is called

    _____ .

9. The main carriers of pollen are _____ and _____ .

10. Insects are attracted to flowers by their nice _____ and bright

    _____ .

WORD SCRAMBLE

Below are several scrambled words you have used in this Lesson. Unscramble the words and write your answers in the spaces provided.

1. TOPLONNILAI   _____

2. SROCS   _____

3. DIWN   _____

4. NISSETC   _____

5. LISTIP   _____

How does fertilization occur?

KEY TERMS

stigma: top part of the pistil

embryo: developing plant

endosperm: tissue that surrounds the developing plant which supplies its food

seed: structure that contains a developing plant and food for its growth

LESSON 29 | How does fertilization occur?

A pollen grain has just landed upon the pistil of a flower. A pollen grain cannot move by itself. How will it get to the ovary to fertilize an egg? The process is very interesting. Check with Figures A through D on the next page as you read how the pollen grain reaches the ovary.

- The top of a pistil has a sticky surface called a **stigma** [STIG-muh]. A pollen grain sticks to the stigma when it lands.

- The nucleus of the pollen grain divides into three parts—one tube nucleus, and two sperm nuclei.

WHAT THE TUBE NUCLEUS DOES

- The tube nucleus <u>dissolves</u> away the material in front of it. It actually <u>digests</u> its way down into the ovary. This makes a path for the two sperm nucleus.

- The two sperm nuclei now move freely down the tube. First they enter the ovary. Then they enter the ovule where the egg is located.

WHAT THE SPERM NUCLEI DO

- One sperm nucleus fertilizes an egg cell. The fertilized egg is called an **embryo** [EM-bree-oh]. The embryo may develop into a new plant.

- The second sperm nucleus joins the rest of the ovule that holds the egg. Together, they become the food the embryo uses when it starts to grow. The stored food for the embryo is called **endosperm** [EN-duh-sperm].

The endosperm and the embryo are surrounded by a tough shell covering. It is formed from the ovule wall. The embryo, the endosperm, and the shell covering together make up a **seed**.

A seed may not grow right away. It can remain dormant for a long time. Something that is dormant is in a resting stage.

When conditions are right, the seed will "wake up." It will sprout and start to grow. During this period, the embryo uses the stored food of the endosperm. When the young plant grows its first leaves, the endosperm is used up. Then the plant makes its own food.

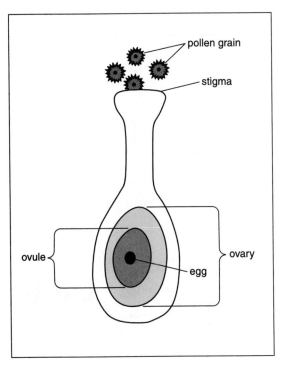

Figure A   *A pollen grain lands on a stigma.*

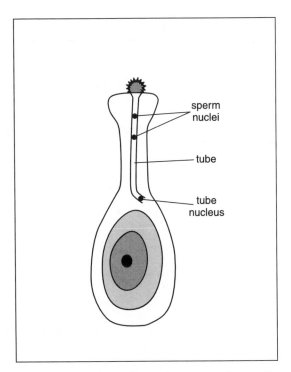

Figure B   *The pollen tube grows down the pistil.*

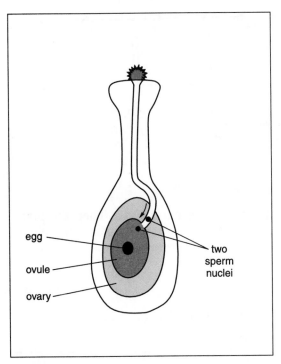

Figure C   *The sperm nuclei enter the ovary.*

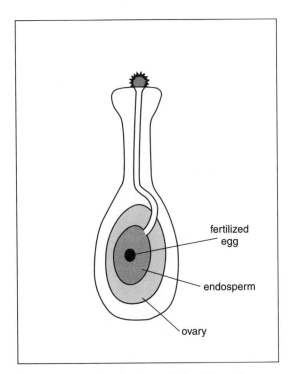

Figure D   *One sperm cell fertilizes the egg. The other joins with the rest of the ovule and becomes the endosperm.*

The embryo and endosperm are surrounded by a covering.

The embryo, endosperm, and covering make a seed.

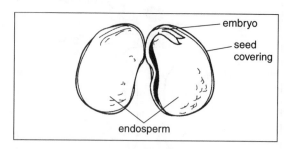

Figure E   *The parts of a seed*

A mature seed will stay dormant until it is planted.

With proper moisture, soil, oxygen, and temperature, the seed will sprout. It will grow to become a new plant.

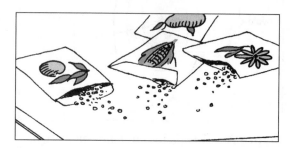

Figure F

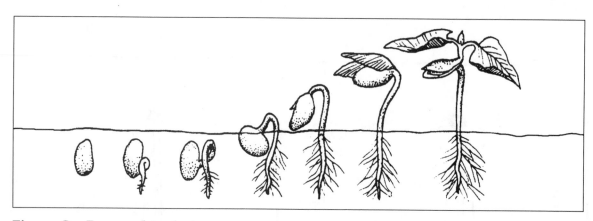

Figure G   *From seed to plant*

The growing embryo feeds on the endosperm until it grows green leaves. Then the plant makes its own food. No matter how you face a seed when you plant it, its stem and leaves grow up and its roots down.

Answer the following questions.

1.   When does a plant embryo stop feeding on the endosperm? _____

_____

2.   What are the three parts of a seed? _____

WHAT DOES THE PICTURE SHOW?

Can you identify the parts of a seed?

Look at Figure H. Identify each part by writing the correct letters in the spaces provided below.

_____ 1. endosperm

_____ 2. seed covering

_____ 3. embryo

Figure H

FILL IN THE BLANK

Complete each statement using a term or terms from the list below. Write your answers in the spaces provided.

dormant   food   stigma
fertilization   embryo   endosperm
pollen grain   seed covering   seed
ovary and ovule

1. In plants, the male gamete is the _____ .

2. The top of the pistil, the _____ , has a sticky surface.

3. The pollen grain moves downward into the _____ .

4. When the sperm and egg combine, _____ takes place.

5. The fertilized egg becomes an _____ .

6. The stored food in a plant is called the _____ .

7. The embryo and stored food are covered by a _____ .

8. The embryo, endosperm, and seed covering make up the _____ .

9. If a seed does not sprout right away it stays _____ .

10. The embryo uses the endosperm until it can make its own _____ .

MATCHING

Match each term in Column A with its description in Column B. Write the correct letter in the space provided.

Column A

_____ **1.** egg

_____ **2.** pollen grain

_____ **3.** embryo

_____ **4.** endosperm

_____ **5.** ovary

Column B

a) young organism

b) male plant gamete

c) food for plant embryo

d) contains ovules

e) female gamete

TRUE OR FALSE

In the space provided, write "true" if the sentence is true. Write "false" if the sentence is false.

_____ **1.** Pollen grains are male gametes.

_____ **2.** Two sperm nuclei fertilize an egg cell.

_____ **3.** A stigma is the top part of a stamen.

_____ **4.** A stigma is slippery.

_____ **5.** An egg is fertilized in an ovule.

_____ **6.** A fertilized plant egg becomes a seed.

_____ **7.** An embryo plant is a full-sized plant.

_____ **8.** The endosperm grows roots and leaves.

WORD SCRAMBLE

Below are several scrambled words you have used in this Lesson. Unscramble the words and write your answers in the spaces provided.

1. IGMSAT   _____

2. GEG   _____

3. MEBYRO   _____

4. IZELIRFET   _____

5. ERMNEDOSP   _____

What is a fruit?

KEY TERM

fruit: swollen plant embryo with one or more ripe seeds

LESSON 30 | What is a fruit?

Is a tomato a vegetable or a fruit? Most people would call it a vegetable. But a tomato is actually a fruit. So is a pepper. So is a cucumber. Does this surprise you?

What is a fruit? Let's start with a definition. A **fruit** is a swollen ovary with one or more ripe seeds.

You have learned that pollen grows into a pistil to fertilize the eggs in the ovary. When the eggs are fertilized, seeds are formed.

Something important happens as the seed or seeds develop. The ovary swells. It swells greatly. In fact the ovary may swell to hundreds or even thousands of times its original size. The large ovary has several important jobs:

- It protects its seed (or seeds) from insects, disease, and bad weather.

- It helps carry the seeds to places where they may grow.

The large ovary and its seeds is called a fruit. Because some fruits taste good, people and animals eat them. There are other fruits that are not good to eat. Some are even poisonous.

So, getting back to the original question — Is a tomato a vegetable or a fruit? A tomato is a ripened ovary and its seeds. So are a pepper and a cucumber. They are all fruits. A vegetable is any other part of a plant that is edible.

A plum tree has flowers in the spring. Then it develops fruit.

Figures A through D show how a plum fruit develops. (Other fruits develop in much the same way.)

Answer the questions after you study the diagrams and read each explanation.

Figure A

Every plum flower has one ovule. Inside the ovule is one egg. A single pollen nucleus fertilizes the egg nucleus.

1. Could a plum flower pollinate itself?

   _____

2. How do you know? _____

   _____

3. In which part of the flower is the ovule

   found? _____

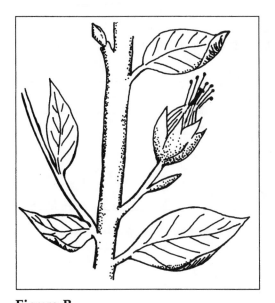

Figure B

Inside the ovary, the fertilized egg divides over and over again. An embryo plant along with its endosperm is forming.

4. What do we call an embryo plant and

   its endosperm? _____

5. What is happening to the ovary as

   the seed is developing? _____

   _____

6. What happened to the petals soon

   after fertilization? _____

   _____

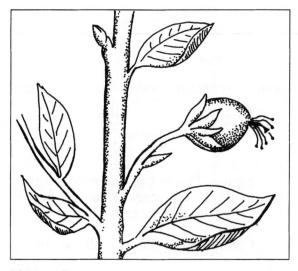

Figure C

The seed continues to develop. The ovary continues to swell.

7. The stamens and the pistil are still on the plant. Can you find them? Draw an arrow to show where they are. Label the arrow.

8. What do you think will happen to

   the stamens and pistil? _____

   _____

   _____

Figure D

The ovary swells even more. The seeds become ripe. A fruit is ripe when the seeds are ready to sprout—or germinate [JER-muh-nate].

9. What do we call a swollen plant

   ovary and its seed (or seeds)?

   _____

10. What has happened to the stamens

    and pistil? _____

    _____

HOW MANY SEEDS DOES A FRUIT HAVE?

The number of seeds depends upon

- how many ovules (eggs) the ovary has.
- how many eggs are fertilized, and
- how many fertilized eggs grow well.

IMPORTANT NOT ALL EGGS ARE FERTILIZED.

NOT ALL FERTILIZED EGGS BECOME SEEDS.

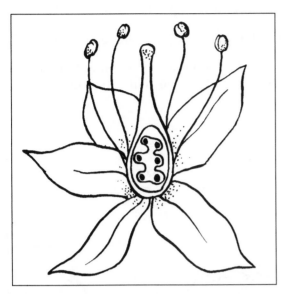

Figure E   *An apple blossom showing its ovary with six ovules*

Figure F   *The ovules produced a fruit with five seeds.*

An ovary with six ovules may become a fruit with six seeds. It may have fewer than six seeds — but no more than six.

Figure E

Answer these questions.

1. Look back at Figure D. How many seeds

    does a plum have? _____

2. Look back at Figure A. How many ovules
    does the ovary of a plum flower have?

    _____

What Do You Think?

3. **a)** Which has more ovules, the ovary of
        a plum or the ovary of a watermelon?

        _____

    b) How do you know? _____

        _____

        _____

203

TYPES OF FRUIT

There are four types of fruit. Two of these types are the fleshy fruits and the dry fruits.

- The ovaries of fleshy fruits fill with water and food material. They are "juicy."

- The ovaries of dry fruits do not fill with water. They are not juicy. Some are very hard.

Several fruits are shown in Figure H. Some are fleshy. Some are dry. Some have many seeds. Some have only one seed. Study the fruits. Then fill in the chart on page 205.

Choose the names of the fruits from the following:

apple	orange	walnut
pear	pea	acorn
peach	peanut	

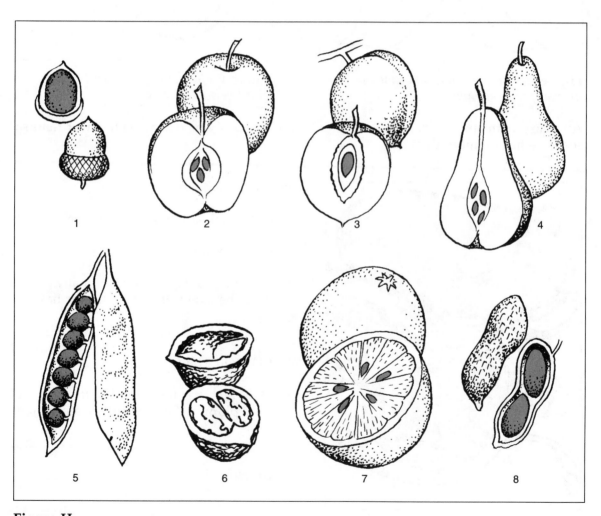

Figure H

	Name of Fruit	Fleshy or Dry?	How Many Seeds?
1.			
2.			
3.			
4.			
5.			
6.			
7.			
8.			

FILL IN THE BLANK

Complete each statement using a term or terms from the list below. Write your answers in the spaces provided.

fleshy	acorn	dry
pistil	swell	ovary
germinate	egg	fruit
peach	pollen grain	

1. The entire female part of a flower is called the _____ .

2. The part of a pistil where ovules are found is the _____ .

3. One ovule produces one _____ .

4. An egg nucleus is fertilized by a single nucleus from a _____ .

5. When an egg is fertilized, its ovary begins to _____ .

6. A swollen plant ovary along with its ripe seeds is called a _____ .

7. A fruit is ripe when its seeds can _____ .

8. Two types of fruit are the _____ fruits and the _____

    fruits.

9. An example of a dry fruit is a _____ .

10. An example of a fleshy fruit is a _____ .

MATCHING

Match each term in Column A with its description in Column B. Write the correct letter in the space provided.

	Column A	Column B
_____	**1.** fruit	**a)** jobs of a fruit
_____	**2.** protects and helps spread its seeds	**b)** types of fruits
_____	**3.** dry and fleshy	**c)** male gamete
_____	**4.** egg	**d)** female gamete
_____	**5.** sperm or pollen	**e)** nature's "seed container"

REACHING OUT

You can now buy some fruits that did not exist before. They are "blends" of two different, but related, fruits.

For example, a <u>tangelo</u> is part tangerine, part grapefruit.

People help nature make this fruit. How do you think this is done?

_____

_____

_____

Figure I

How are seed dispersed?

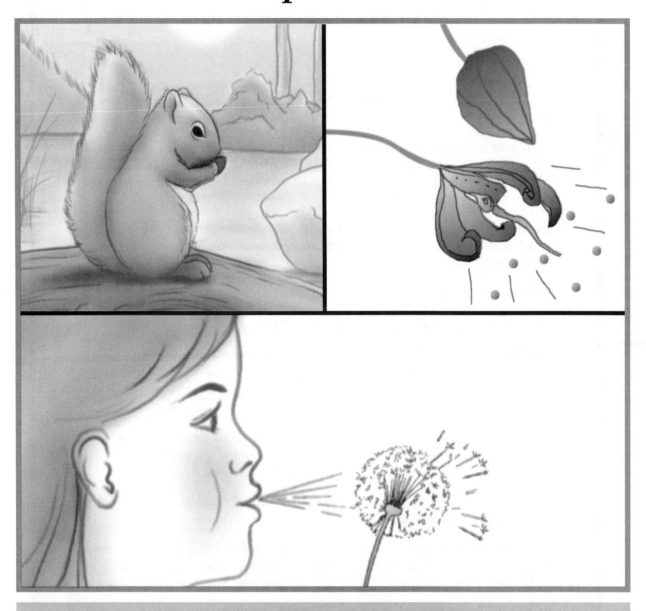

KEY TERM

seed dispersal: spreading of seeds

LESSON 31 | How are seeds dispersed?

Did you ever hear anyone say, "The apple doesn't fall far from the tree?" Sometimes fruits containing seeds drop close to the parent plant.

Can you imagine what would happen if all seeds were to fall near their parent plants? Too many plants would start to grow in one place. Soon there would not be enough soil, water, oxygen, and sunlight for all of them. Only a few would live.

Seed-spreading is needed for plants to survive. Seed-spreading is called **seed dispersal**. Seeds are dispersed in several ways. These are the most important:

WIND Some plants have fruits that are light and shaped for being carried by the wind. For example, the fruit of maple trees have "wings." The fruits of dandelions have parachute-like tufts.

ANIMALS Animals eat many kinds of fruits, seeds and all. Many kinds of seeds are not changed by digestion. They remain "healthy" and can still germinate. As the animals move from place to place, they get rid of the seeds as waste. This is one way animals disperse seeds.

Animals disperse seeds in other ways too. The seeds of some fruits have sticky parts. For example, the cocklebur has hooks. They stick to the fur of passing animals. Then they fall off somewhere else.

Some animals, like squirrels, carry off and bury nuts for the winter. Later on, some of the nuts germinate.

EXPLOSIVE PARTS The ovaries of some fruits—like the poppy and the pea—"explode" when they mature. The seeds shoot out and land at a distance from the plant.

WATER Some fruits that grow near water can float. When the fruits fall in the water, currents carry them to far away land areas. Coconut and lotus seeds are dispersed by water.

Study Figures A through G and answer the questions about each picture.

Figure A

Maple, milkweed, and dandelion fruits have "helpers" to fly about. Why do tiny fruits not need "helpers"?

Figure B

This bird is eating berries—seeds and all. What will happen to the seeds?

Figure C

In time, what will happen to the cockle-bur fruits sticking to this dog?

Figure D

What will happen to the seeds of the acorn this squirrel is holding?

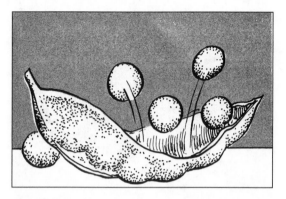

When a pea pod dries, it twists. The pod breaks open suddenly. Out shoot the peas.

Figure E

Figure F

These fruits shoot their seeds into the air, too.

Figure G

Coconuts are lighter than water. How is this helpful to their seed dispersal? _____

FILL IN THE BLANK

Complete each statement using a term or terms from the list below. Write your answers in the spaces provided.

seed dispersal	oxygen	cocklebur
"explodes"	soil	sunlight
wind	bury	water
germinate	dandelions	maple trees
digestion	water currents	survive

1. Too many plants cannot grow in one small place because they all cannot get enough

   _____, _____, _____, and _____ .

2. The spreading of seeds is called _____ .

3. Many kinds of seeds are carried through the air by the _____ .

4. Two plants with seeds that are carried by wind are _____ and

   _____ .

5. A _____ has hooks that stick to animals.

6. Many kinds of seeds are not harmed by _____ .

7. A pea pod _____ its seeds out when it matures.

8. Squirrels sometimes _____ nuts. Some of them may become ripe and

   _____ .

9. Coconut seeds are dispersed great distances by _____ .

10. Seed dispersal is needed for plants to _____ .

MATCHING

Match each term in Column A with its description in Column B. Write the correct letter in the space provided.

	Column A	Column B
_____	1. seed dispersal	a) dispersed by water
_____	2. coconut seeds	b) does not harm many kinds of seeds
_____	3. dandelion seeds	c) dispersed by wind
_____	4. squirrels	d) any seed-spreading
_____	5. digestion	e) help disperse acorns

WORD SCRAMBLE

Below are several scrambled words you have used in this Lesson. Unscramble the words and write your answers in the spaces provided.

1. TERAW _____

2. PLOSEXIVE _____

3. MALSINA _____

4. SPERISDAL _____

5. IDNW _____

REACHING OUT

Figure H

Many species of plants produce many more seeds than is needed for the species to survive. Why are so many seeds produced? _____

_____

_____

What are tropisms?

KEY TERMS

tropism: response of plants to stimuli

phototropism: response of a plant to light

geotropism: response of a plant to gravity

hydrotropism: response of a plant to water

thigmotropism: response of a plant to touch

LESSON 32 | What are tropisms?

Plants do not move from place to place the way animals do. However, all plants respond to stimuli. A stimulus is a change in the environment that causes a response. The response of plants to stimuli are called **tropisms** [TROH-pizmz]. A plant responds to a stimulus by growing in a certain direction.

LIGHT

All plants grow toward light. Have you ever noticed that the stem and leaves of a plant grow toward the sun? The response of a plant to sunlight is called **phototropism** [foh-TAH-troh-piz-um]. Photo- means light.

In some plants, the petals of its flower open in the morning when the sun rises. When the sun sets, the petals of the flower close again. The petals react to the light that comes from the sun.

GRAVITY

Did you ever wonder why farmers do not worry about which way to plant seeds in the ground? Farmers need not worry because the roots of all plants grow downward and the stems upward. The direction of the growth of roots and stems is the response of a plant to gravity. Roots grow down in response to gravity. Stems grow up. The response of a plant to gravity is called **geotropism** [jee-AH-truh-piz-um] Geo- means earth.

WATER

A plant's roots also grow toward water. The growth of a plant's roots toward water is called **hydrotropism** [hy-DRAHT-ruh-pizm]. Hydro- means water. In most plants, this tropism is not very strong. It occurs only when water touches the roots.

TOUCH

Some plants respond when they are touched. For example, the mimosa plant closes up its leaves when they are touched. Grapevines often coil around fence posts or climb up walls. The response of a plant to touch is called **thigmotropism** [thig-MAH-truh-piz-um]. Thigmo- means touch.

WHAT DO THE PICTURES SHOW?

Each picture shows a plant tropism. Figure out what tropism is shown in each picture. Write your answers on the lines under each picture.

Figure A

1. TROPISM?_____

Figure B

2. TROPISM?_____

Figure C

3. TROPISM?_____

Figure D

4. TROPISM?_____

TRY THIS EXPERIMENT

Get a potted plant.

Place the plant near a window that receives sunlight.

Place the plant so that its leaves and/or flowers are facing in a direction away from the light.

Observe the plant each day for the next two weeks.

Figure E

CONCLUSIONS

1. What happened to the position of the plant's leaves and flowers? _____

   _____

2. What kind of tropism is this? _____

3. What does the term <u>photo-</u> mean? _____

AN UNUSUAL PLANT

Some plants respond immediately to stimuli. When an insect touches the leaves of a Venus' flytrap, the leaves of the plant snap shut around the insect. The insect is then trapped in the plant's leaves.

Special chemicals are given off by the leaf. The chemicals digest the soft parts of the insect. After the insect has been digested, the trap opens again.

Figure F

TRUE OR FALSE

In the space provided, write "true" if the sentence is true. Write "false" if the sentence is false.

_____ 1. The response of a plant to light is called phototropism.

_____ 2. The petals of a flower open when the sun rises.

_____ 3. A plant's roots grow away from water.

_____ 4. <u>Geo-</u> means earth.

_____ 5. The leaves of a mimosa plant open when they are touched.

_____ 6. In most plants, hydrotropism is very strong.

_____ 7. Roots grow down in response to gravity.

_____ 8. A Venus' flytrap responds immediately to touch.

_____ 9. <u>Thigmo-</u> means water.

_____ 10. All plants grow towards light.

WORD SCRAMBLE

Below are several scrambled words you have used in this Lesson. Unscramble the words and write your answers in the spaces provided.

1. ARETW   _____

2. PMOTRIS   _____

3. GILTH   _____

4. OTROS   _____

5. RAGTYIV   _____

SCIENCE *EXTRA*

Landscape Architect

Do you enjoy growing plants and flowers? Does a career that mixes art and science interest you? If you enjoy working inside as well as outside, a career as a landscape architect may be for you. A landscape architect is just one of the many careers available in the horticulture [HOWR-tuh-kul-chur] field. Horticulture is the study of plants.

Landscape architects design and develop land for human use and enjoyment. They are concerned with the environment and must plan the best and most practical use of the land.

Landscape architects often create beautiful outdoor environments. Their projects range from large to small. A landscape architect may design an entire city park or the layout of a new golf course. They also may design something as small as a single-family garden.

Landscape architects also work indoors. They design the landscaped gardens or plant areas often found in office buildings, shopping malls, and hotel lobbies.

Landscape architects get involved in all the stages of planning. They study the area that they have been hired to landscape. They study the climate, temperature, water supply, and vegetation of the area. Then, they pick the plants or flowers that would best grow in this environment.

Indoor landscape architects also may work with building architects. Together, they try to develop designs that will best take advantage of the sunlight and water supply that might be available.

To become a landscape architect, you should enjoy taking science classes, especially biology. It also is helpful to have artistic ability and an "eye" for design. Many community colleges offer a two-year program in horticulture. Many people continue their studies for advanced degrees.

What are invertebrates?

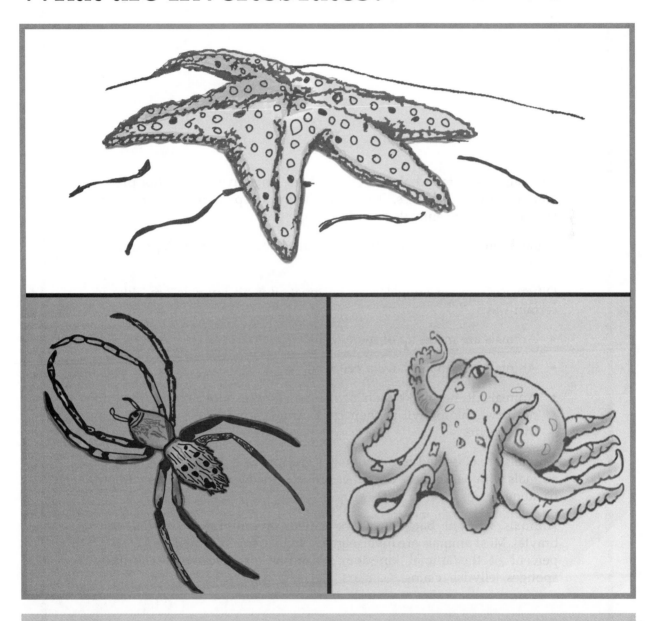

KEY TERM

invertebrates: animals without backbones

LESSON 33 | What are invertebrates?

Animals can be found almost everywhere. They live in very hot places, and in very cold places. They live in the deep ocean, and on high mountains. Some animals even live in the ground.

Some animals are small. Others, like the whale and the elephant, are large.

Differences among animals can be great. But all animals are alike in certain ways.

- Animals are made up of many cells.

- Animal cells do not have a cell wall.

- Animals do not make their own food. They cannot because they do not have chlorophyll in their cells. Animals must take in food from the outside.

Animals are classified into two major groups. One group is made up of animals with backbones. The other group is made up of animals without backbones.

Animals without backbones are called **invertebrates** [in-VUR-tuh-brayts]. Most animals are invertebrates. In fact, invertebrates make up 97 percent of the animal kingdom. Examples of invertebrates include sponges, jellyfish, clams, sea stars (starfish), worms, and insects.

CLASSIFICATION OF INVERTEBRATES

COMMON NAME	EXAMPLES		IMPORTANT CHARACTERSTICS
SPONGES			• saclike bodies • most live attached to objects on the ocean floor • have many pores (holes) through which water flows
CNIDARIANS [ni-DER-ee-uns]	jellyfish corals hydra		• have tentacles • all have stinging cells • live in water
FLATWORMS	tapeworms flukes planaria		• long, flat, ribbonlike bodies • some get food by living in another organism and absorbing food from that organism
ROUNDWORMS	hookworms		• long, thin, tubelike bodies • some get food by living in another organism and absorbing food from that organism
SEGMENTED WORMS	earthworms leeches		• long tubelike body that is divided into segments (sections) • simplest organisms with a well-developed nervous system
MOLLUSKS [MAHL-usks}	snails clams squids		• soft bodies • many have shells • most live in the ocean; some live in fresh water and on land
ECHINODERMS [ee-KY-noh-durms]	sea stars sand dollars sea cucumbers		• usually have five arms that extend from a middle body section • have an internal skeleton made up of spines • live only in the ocean
ARTHROPODS [ar-thruh-PODS]	spiders lobsters centipedes grasshoppers		• have jointed legs • have a hard outer covering • have segmented bodies • live on land and in water

What do vertebrates have that invertebrates lack?

Study Figures A through H and the chart on Page 221. Then answer the questions.

Figure A

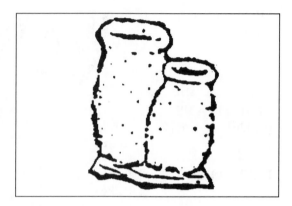

Figure B

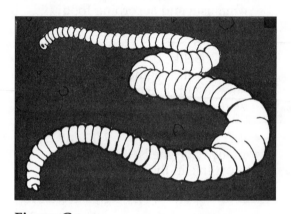

Figure C

Figure D

Figure E

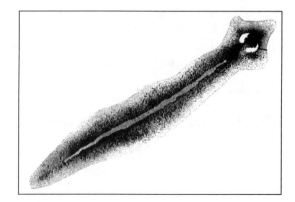

Figure F

Figure G

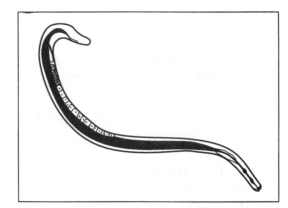

Figure H

1. Do the animals shown have a backbone? _____
   <small>yes, no</small>

2. Do these animals take in food from the outside? _____
   <small>yes, no</small>

3. Are these animals made up of many cells? _____
   <small>yes, no</small>

4. Do their cells have a cell wall? _____
   <small>yes, no</small>

5. Do their cells contain chlorophyll? _____
   <small>yes, no</small>

6. These animals are _____ .
   <small>vertebrates, invertebrates</small>

7. Are most animals invertebrates? _____
   <small>yes, no</small>

8. Which of the animals shown is the simplest organism with a well-developed

   nervous system? _____

9. Which is a mollusk? _____

10. Which animal has jointed legs? _____

11. Which animal has stinging cells? _____

12. Which animal has a long, flat, ribbonlike body? _____

13. Which animal has many pores? _____

14. Which animal has a long, tubelike body that is *not* divided into segments?

    _____

15. Which animal is an echinoderm? _____

ABOUT ARTHROPODS

Arthropods make up the largest phylum in the animal kingdom. In addition to insects, this phylum also includes animals such as lobsters, spiders and ticks, centipedes, and millipedes.

Look at the chart below. The chart lists major groups of arthropods. It also lists their important characteristics.

MAJOR GROUP	IMPORTANT CHARACTERISTICS
Centipedes	Centipedes have flat bodies with segments. Each body segment has one pair of legs.
Millipedes	Millipedes have round bodies with many segments. Each body segment, except the first four, has two pairs of legs.
Crustaceans [krus-TAY-shuns] Includes: lobsters, shrimp, crabs, and crayfish	Crustaceans have two main body parts, two pairs of antennae, large claws, and four pairs of legs. Most crustaceans live in water.
Arachnids [uh-RAK-nids] Includes: spiders, scorpions, and ticks	Arachnids have two body sections and four pairs of legs.
Insects	Insects have three body sections and three pairs of legs. Some insects have wings.

MATCHING

Match each term in Column A with its description in Column B. Write the correct letter in the space provided.

Column A	Column B
_____ 1. arachnids	**a)** have three body sections
_____ 2. centipedes	**b)** most live in water
_____ 3. millipedes	**c)** have round bodies with many segments
_____ 4. insects	**d)** includes ticks
_____ 5. crustaceans	**e)** have flat bodies with many segments

CLASSIFY THE ARTHROPOD

Look at the pictures of the arthropods below. What arthropod group does each organism belong to? Write the name of the group on the line below each picture.

Figure I

1. _____

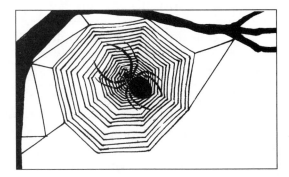

Figure J

2. _____

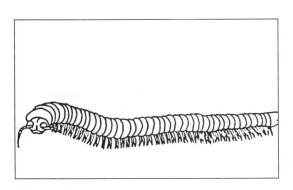

Figure K

3. _____

Figure L

4. _____

Figure M

5. _____

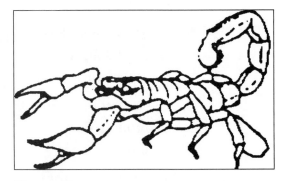

Figure N

6. _____

PLANT OR ANIMAL?

Complete the chart below by putting a check in the correct box or boxes for each characteristic.

	CHARACTERISTIC	PLANTS	ANIMALS
1.	living organisms		
2.	cells do not have chlorophyll		
3.	cells have no cell wall		
4.	make their own food		
5.	cells have chlorophyll		
6.	are made up of many cells		
7.	cells have a cell wall		
8.	take in food from the outside		
9.	most move about from place to place		
10.	carry out every life process		

MULTIPLE CHOICE

In the space provided, write the letter of the word that best completes each statement.

_____ **1.** Most animals are
  a) protists.   **b)** vertebrates.
  c) invertebrates.   **d)** sponges.

_____ **2.** Examples of invertebrates include
  a) birds.   **b)** insects.
  c) rabbits.   **d)** fish.

_____ **3.** Sea stars are in the same group of invertebrates as
  a) sand dollars.   **b)** grasshoppers.
  c) tapeworms.   **d)** jellyfish.

_____ **4.** The largest phylum in the animal kingdom is made up of
  a) roundworms.   **b)** sponges.
  c) mullusks.   **d)** arthropods.

_____ **5.** Snails and clams are
  a) sponges.   **b)** planaria.
  c) mollusks.   **d)** cnidarians.

How do insects develop?

KEY TERMS

metamorphosis: changes during their stages of development of an insect

larva: wormlike stage of insect development

pupa: resting stage during complete metamorphosis

cocoon: protective covering around the pupa

nymph: young insect that looks like the adult

LESSON 34 | How do insects develop?

Insects are by far the largest single group of invertebrates. Nearly one million insect species have been classified so far, and every year, scientists discover thousands of more insect species. Some scientists think that there are as many as 10 million insect species. And, each species contains billions of members. Can you imagine how many insects there are all together?

You have learned about many of the insect characteristics in Lesson 33. Now you will learn how insects develop.

All insects hatch from eggs. When an insect hatches from its egg, the insect usually does not look like the adult insect. Insects grow into adults by **metamorphosis** [met-uh-MOR-fuh-sis]. Metamorphosis means a change into a new form. There are two types of metamorphosis, complete and incomplete.

COMPLETE METAMORPHOSIS

Did you ever see a butterfly among the flowers? Just a short time before, it was a crawling caterpillar. The caterpillar changes form completely to become a butterfly. It underwent complete metamorphosis.

Some insects that develop by complete metamorphosis are butterflies, mosquitos, flies, and bees. These insects have four stages in their development: egg, **larva** [LAHR-vuh], **pupa** [PYOU-puh], and adult. The egg hatches into a larva (such as a caterpillar). Then the larva becomes a pupa. During this stage, many insects form a **cocoon** [kuh-KOON]. Finally out comes the adult. The adult looks nothing like the larva. The adults mate and the cycle starts all over again.

INCOMPLETE METAMORPHOSIS

During incomplete metamorphosis, the change from the young stage to the adult stage is not so great. There are only three stages of development: egg, **nymph** [NIMF], and adult.

Grasshoppers develop by incomplete metamorphosis. Grasshopper eggs hatch into nymphs. A **nymph** is a young grasshopper. A nymph looks like an adult only smaller. It has a large head but no wings. The wings develop as the nymph grows.

A nymph sheds or molts its outer covering about five times as it grows. Small wings appear after the first molt. The wings grow larger with each molt. The last molt produces an adult grasshopper with full wings.

BUTTERFLY DEVELOPMENT

Study Figure A. Then answer the questions.

COMPLETE METAMORPHOSIS

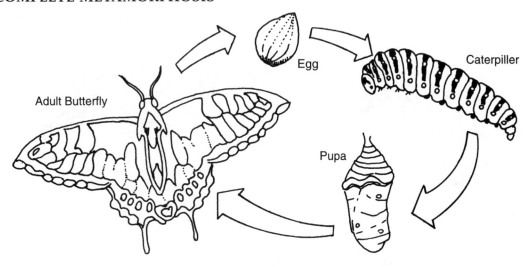

Figure A

1. What kind of animal is the butterfly? _____

2. Butterflies reproduce by _____ reproduction.
   _{sexual, asexual}

3. Where does a female butterfly lay her fertilized eggs? _____

4. Some animals change in form as they develop into adults. What is this change called?

   _____

5. **a)** Do butterflies develop by metamorphosis? _____

   b) What kind? _____

   c) List the stages of this development. _____ _____

   _____ _____

6. Does the larva stage resemble the adults? _____

7. Does the pupa stage resemble the adult? _____

8. What is the common name of a butterfly's pupa stage? _____

9. What is the common name of a butterfly's larva stage? _____

10. A butterfly molts just once.

    a) From which stage does a butterfly molt? _____

    b) Into which stage does a butterfly molt? _____

GRASSHOPPER DEVELOPMENT

Study Figure B. Then answer the questions.

INCOMPLETE METAMORPHOSIS

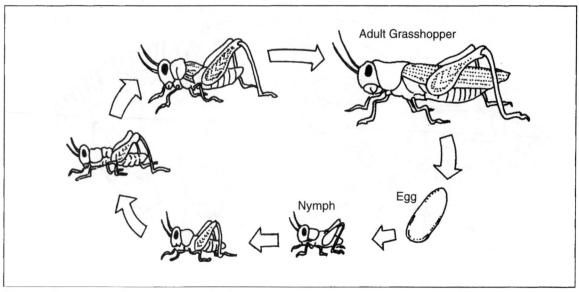

Figure B

1. What kind of animal is the grasshopper? _____

2. Grasshoppers reproduce by _____ reproduction.
   <small>sexual, asexual</small>

3. Where does a female grasshopper lay her fertilized eggs? _____

4. **a)** Do grasshoppers develop by metamorphosis? _____

   b) What kind? _____

   c) List the stages of incomplete metamorphosis. _____

   _____   _____

5. Does the nymph stage resemble the adult? _____

6. A nymph is _____ than an adult.

7. A nymph has no _____ .

8. A nymph has a _____ head.
   <small>big, small</small>

9. A nymph sheds its outer covering several times. What is another word for shed?

   _____

10. A nymph grasshopper grows in size. What appears after a nymph sheds its covering

    the first time? _____

FILL IN THE BLANK

Complete each statement using a term or terms from the list below. Write your answers in the spaces provided.

head	adult	insects	complete
do not	incomplete	caterpillar	wings
metamorphosis	cocoon	eggs	
molts	nymphs		

1. During the pupa stage, many insects form a _____ .

2. Insects hatch from _____ .

3. Insects change in form as they grow. This is called _____ .

4. Egg, larva, pupa, and adult are the stages of _____ metamorphosis.

5. Larva and pupa stages _____ resemble the adult.

6. Egg, nymph, and adult are the stages of _____ metamorphosis.

7. A nymph resembles the _____ .

8. A nymph _____ about five times.

9. A nymph has a large _____ , but no _____ .

10. _____ are the largest single group of invertebrates.

11. Grasshopper eggs hatch into _____ .

12. The larva stage of a butterfly is a _____ .

MATCHING

Match each term in Column A with its description in Column B. Write the correct letter in the space provided.

	Column A	Column B
_____	1. molt	a) a butterfly's larva stage
_____	2. caterpillar	b) young grasshopper
_____	3. nymph	c) stages of incomplete metamorphosis
_____	4. egg, nymph, adult	d) shed outer covering
_____	5. egg, larva, pupa, adult	e) stages of complete metamorphosis

COMPLETE THE CHART

Complete the chart by writing the stage of development for each illustration. Then number each stage in the correct order of development from A to D.

		Stage	Order of Development
1. 			
2. 			
3. 			
4. 			

WORD SCRAMBLE

Below are several scrambled words you have used in this Lesson. Unscramble the words and write your answers in the spaces provided.

1. MORSHOPMETSIA   _____

2. MPELETCO   _____

3. VARAL   _____

4. LUDAT   _____

5. PHYNM   _____

What are vertebrates?

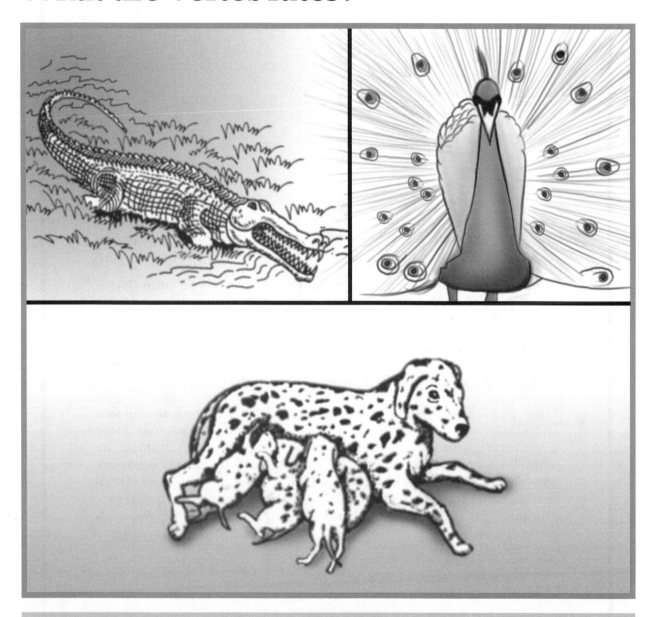

KEY TERMS

vertebrate: animals with a backbone

endoskeleton: internal skeleton

vertebra: small bones that make up the backbone

LESSON 35 | What are vertebrates?

What do goldfish, frogs, turtles, robins, and people all have in common? They are **vertebrates** [VUR-tuh-brits]. Vertebrates are one of the two major groups of animals. What is the other?

Vertebrates are animals with a backbone. The backbone protects the spinal cord. The spinal cord is made up of many nerves. All vertebrates have a nerve cord. The backbone also helps support an animal.

Vertebrates also have a skeleton inside their bodies. The internal skeleton is called an **endoskeleton** [en-duh-SKEL-uh-tun]. The endoskeleton protects the inside organs and gives support to the body.

There are five major groups of vertebrates. They are fishes, amphibians [am-FIB-ee-unz], reptiles, birds, and mammals. You can probably name some kinds of fishes and birds. But, do you know what animals are amphibians, reptiles, or mammals? Salamanders, frogs, and toads are amphibians. Two kinds of reptiles are crocodiles and alligators. Dinosaurs were reptiles too. Kangaroos, dogs, cats, elephants, and gorillas are just some of the many different kinds of mammals. What other mammals can you name?

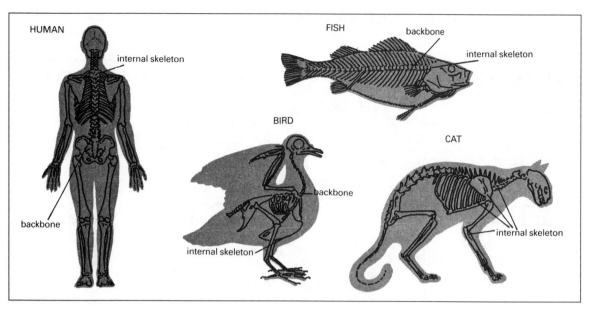

Figure A

1. Look at Figure A. These animals are _____ .
   vertebrates, invertebrates

2. What does each of these animals have running down the middle of its back?

   _____

3. The backbone is part of an internal _____ .

4. Any animal with a backbone and an internal skeleton is classified as

   _____ .
   a vertebrate, an invertebrate

5. The backbone protects the _____ .

NOW TRY THIS

You can feel your backbone.

To find your backbone, run your fingers up the center of your back. Each bump that you feel is a **vertebra** [VUR-tuh-brah]. The backbone is made up of small bones called vertebrae.

How far up and down your back does your

backbone go? _____

_____

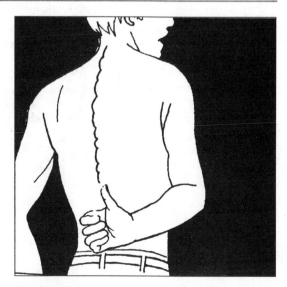

Figure B

CLASSIFICATION OF VERTEBRATES

COMMON NAME	EXAMPLES	IMPORTANT CHARACTERISTICS
FISHES		• live entirely in water • obtain dissolved oxygen from water through gills • cold-blooded • most lay eggs • external fertilization
AMPHIBIANS		• live part of their lives in water and part on land • undergo metamorphosis (see Lesson 18) • water-dwellers breathe with gills; land dwellers breathe with lungs • cold-blooded • smooth, moist skin • lay eggs • most have external fertilization
REPTILES		• live mainly on land • breathe air with lungs • cold-blooded • dry skin covered with hard plates or scales • most lay eggs which are hard-shelled • internal fertilization
BIRDS		• live on land • breathe air with lungs • warm-blooded • body covered with feathers • have lightweight bones • lay hard-shelled eggs • internal fertilization
MAMMALS		• most live on land • breathe air with lungs • warm-blooded • body covered with fur or hair • in most, young develop entirely inside the mother's body • produce milk to nurse their young

Use what you have read about vertebrates and the chart on page 236 to answer the following questions about vertebrates.

1. Which groups of vertebrates have a backbone? _____

_____ _____

2. What are two jobs of the backbone? _____

_____

3. The internal skeleton of a vertebrate is called an _____ .
exoskeleton, endoskeleton

4. What are the five major groups of vertebrates? _____

_____

5. Which are the only vertebrates that

   a) have hair? _____

   b) have feathers? _____

   c) can breathe on land and in water? _____

   d) feed milk to their babies? _____

6. Which vertebrates are

   a) warm-blooded? _____

   b) cold-blooded? _____

7. Which vertebrates breathe only in water through gills? _____

8. Which vertebrates have

   a) internal fertilization? _____

   b) external fertilization? _____

9. Which vertebrates are covered with hard plates? _____

10. In which vertebrates do most young develop inside the mother's body?

_____

Today mammals are one of the most successful groups of animals on earth. In fact, the present time often is referred to as the Age of Mammals.

Scientists classify mammals into three major groups. Mammals are grouped according to the way in which their young develop.

Egg-Laying Mammals

You probably did not think that mammals lay eggs, but some do. The platypus [PLAT-uh-pus] and the spiny anteater are two kinds of mammals that lay eggs. Their young develop outside the mother's body inside shell-covered eggs.

Figure C   *Egg-Laying Mammals*

Pouched Mammals

In pouched mammals, the young are born at an early stage of development. They are not fully developed. The young complete their development in a pouch on the mother. Kangaroos, opossums, and koala bears are pouched mammals.

Figure D   *Pouched Mammals*

Placental Mammals

The largest group of mammals is the placental [pluh-SEN-tul] mammals. These mammals give birth to young that develop completely inside the mother's body. Why are these mammals called placentals? The young are connected to the mother by a saclike organ called the placenta [pluh-SEN-tuh]. The placenta gives the developing mammal food. It also carries away wastes.

Figure E   *Placental Mammals*

MORE ABOUT MAMMALS

Look at the chart below. The chart lists the major groups of placental mammals. It also lists the names of some animals that belong to each group.

COMMON NAME	EXAMPLES
Insect-eating mammals	moles, shrews
Flying mammals	bats
Toothless mammals	armadillos, anteaters, sloths
Gnawing mammals	rodents, such as mice and beavers
Rodentlike mammals	rabbits, hares
Aquatic mammals	whales, dolphins, porpoises, sea cows
Trunk-nosed mammals	African elephants, Asian elephants
Carnivorous [KAR-nuh-vaw-rus] mammals	dogs, bears, seals, walruses
Hoofed mammals	horses, cows, camels
Primates	monkeys, apes, humans

TRUE OR FALSE

In the space provided, write "true" if the sentence is true. Write "false" if the sentence is false.

_____ **1.** Pouched mammals lay eggs.

_____ **2.** Moles eat plants.

_____ **3.** Bats are mammals that can fly.

_____ **4.** The largest group of mammals is the egg-laying mammals.

_____ **5.** Horses and camels are hoofed mammals.

_____ **6.** Aquatic mammals breathe through gills.

_____ **7.** Apes and humans are primates.

_____ **8.** The placenta is a saclike organ.

_____ **9.** People are placental mammals.

_____ **10.** Kangaroos and opossums are pouched mammals.

METRIC-ENGLISH CONVERSIONS

	Metric to English	*English to Metric*
Length	1 kilometer = 0.621 mile (mi)	1 mi = 1.61 km
	1 meter = 3.28 feet (ft)	1 ft = 0.305 m
	1 centimeter = 0.394 inch (in)	1 in = 2.54 cm
Area	1 square meter = 10.763 square feet	1 ft^2 = 0.0929 m^2
	1 square centimeter = 0.155 square inch	1 in^2 = 6.452 cm^2
Volume	1 cubic meter = 35.315 cubic feet	1 ft^3 = 0.0283 m^3
	1 cubic centimeter = 0.0610 cubic inches	1 in^3 = 16.39 cm^3
	1 liter = .2642 gallon (gal)	1 gal = 3.79 L
	1 liter = 1.06 quart (qt)	1 qt = 0.94 L
Mass	1 kilogram = 2.205 pound (lb)	1 lb = 0.4536 kg
	1 gram = 0.0353 ounce (oz)	1 oz = 28.35 g
Temperature	Celsius = 5/9 (°F –32)	Fahrenheit = 9/5°C + 32
	0°C = 32°F (Freezing point of water)	72°F = 22°C (Room temperature)
	100°C = 212°F (Boiling point of water)	98.6°F = 37°C (Human body temperature)

METRIC UNITS

The basic unit is printed in capital letters.

Length	*Symbol*
Kilometer	km
METER	m
centimeter	cm
millimeter	mm

Area	*Symbol*
square kilometer	km^2
SQUARE METER	m^2
square millimeter	mm^2

Volume	*Symbol*
CUBIC METER	m^3
cubic millimeter	mm^3
liter	L
milliliter	mL

Mass	*Symbol*
KILOGRAM	kg
gram	g

Temperature	*Symbol*
degree Celsius	°C

SOME COMMON METRIC PREFIXES

Prefix		*Meaning*
micro-	=	0.000001, or 1/1,000,000
milli-	=	0.001, or 1/1,000
centi-	=	0.01, or 1/100
deci-	=	0.1, or 1/10
deka-	=	10
hecto-	=	100
kilo-	=	1,000
mega-	=	1,000,000

SOME METRIC RELATIONSHIPS

Unit	*Relationship*
kilometer	1 km = 1,000 m
meter	1 m = 100 cm
centimeter	1 cm = 10 mm
millimeter	1 mm = 0.1 cm
liter	1 L = 1,000 mL
milliliter	1 mL = 0.001 L
tonne	1 t = 1,000 kg
kilogram	1 kg = 1,000 g
gram	1 g = 1,000 mg
centigram	1 cg = 10 mg
milligram	1 mg = 0.001 g

GLOSSARY/INDEX

adaptation [ad-up-TAY-shun]: characteristic of an organism that helps the organism survive, 19

algae [AL-jee]: plant-like group of protists that can make their own food, 147

amphibian [am-FIB-ee-un]: animal that lives part of its life in water and part on land, 117

anther: structure of a flower that produces pollen grains, 181

asexual reproduction: kind of reproduction that requires only one parent, 65

binary fission [BY-nur-ee FIZH-un]: form of asexual reproduction in which one cell divides into two identical cells, 77

budding: form of asexual reproduction in which a small part of a cell breaks off to form a new organism, 77

bulb: underground stem with fleshy leaves, 95

caustic: able to burn and irritate the skin, 15

cell: basic unit of structure and function in all living things, 41

cell division: process by which cells reproduce, 71

cell membrane: thin "skin" that covers the cell and gives the cell its shape, 41

cellulose [SEL-yoo-lohs]: nonliving substance that makes up plant cell walls, 159

chitin [KYT-in]: hard substance that makes up the cell walls of fungi, 153

chlorophyll [KLAWR-uh-fill]: green pigment needed by a plant in order to make its own food, 159

chloroplast [KLAWR-uh-plast]: structure in the cells of a green plant that store chlorophyll, 159

circulation [sur-kyuh-LAY-shun]: movement of products throughout an organism, 25

cocoon [kuh-KOON]: protective covering around the pupa, 227

compound leaf: leaf that has its leaf blade divided into smaller leaflets, 167

cross-pollination: when pollen is carried from the stamen of a flower on one plant to the pistil of a flower on a different plant, 187

cytoplasm [SYT-uh-plaz-um]: living material inside the cell membrane, excluding the nucleus, 41

data [DAYT-uh]: record of observations, 9

digestion [dy-JES-chun]: breaking down of food into usable forms, 25

embryo [EM-bree-oh]: developing organism, 193

endosperm [EN-duh-spurm]: tissue that surrounds the developing plant which supplies its food, 193

excretion [ik-SKREE-shun]: getting rid of waste products, 25

fertilization [fur-tul-i-ZAY-shun]: union of one sperm cell and an egg cell, 103

flagella [fluh-JEL-uh]: hairlike structures that bacteria use to move, 139

fruit: swollen plant embryo with one or more ripe seeds, 199

fungi: plantlike organisms that lack chlorophyll, 153

gametes [GAM-eets]: reproductive cells, 125

genus [JEE-nus]: classification group made up of related species, 131

gills: organs that absorb dissolved oxygen from water, 117; structures that produce mushroom spores, 153

hypothesis [hy-PAHTH-uh-sis]: suggested solution to a problem based upon known information, 9

imperfect: flower that has only pistils or stamens, 181

ingestion [in-JES-chun]: taking in of food, 25

invertebrates [in-VUR-tuh-brayts]: animals without backbones, 219

kingdom: largest classification group, 131

larva [LAHR-vuh]: wormlike stage of insect development, 227

length: distance between two points, 1

mass: amount of matter in an object, 1

meiosis [my-OH-sis]: process by which gametes form, 125

metamorphosis [met-uh-MOWR-fuh-sis]: changes during the stages of development of an organism, 117, 227

241

microscope: instrument that makes things appear larger than they really are, 49

mitosis [my-TOH-sis]: kind of cell division in which the nucleus divides, 71

monerans [muh-NER-uns]: single-celled organisms that do not have a nucleus, 139

nucleus [NEW-klee-us]: part of a cell that controls the cell's activities, 41

nymph [NIMF]: young insect that looks like the adult, 227

organism [AWR-guh-nizm]: living things, 19

ovule [OH-vyool]: small structures in which egg cells develop in a flower, 181

perfect flower: flower that has both pistils and stamens, 181

photosynthesis [foht-uh-SIN-thuh-sis]: food-making process in plants, 175

phylum [FY-lum]: classification group made up of related classes, 131

pistil: female reproductive organ of a flower, 181

pollen: male plant reproductive cell, 181

pollination [pahl-uh-NAY-shun]: movement of pollen from a stamen to a pistil, 187

protist: simple organisms that have a true nucleus, 147

protozoan [proht-uh-ZOH-un]: one celled animal-like protist that cannot make its own food, 147

pupa [PYOU-puh]: resting stage during complete metamorphosis, 227

regeneration [ri-jen-uh-RAY-shun]: ability of an animal to regrow lost body parts, 89

reproduction [ree-pruh-DUK-shun]: life process by which organisms produce new organisms, 65

respiration [res-puh-RAY-shun]: process by which organisms release energy from food, 25, 175

response: reaction to a change in the environment, 19

rhizoids [RY-zoidz]: rootlike structures that anchor fungi, 153

root cap: covers and protects the tip of a root, 167

root hairs: tiny hairlike structures that help a root absorb more water, 167

safety alert symbols: signs that warn of hazards or dangers, 15

scientific method: problem solving guide, 9

seed: reproductive structure, 159; structure that contains a developing plant and food for its growth, 193

seed dispersal: spreading of seeds, 207

self-pollination: when pollen is carried from the stamen of one flower to the pistil of another flower on the same plant, 187

sepal [see-PUHL]: special leaves that support and protect the flower bud, 181

sexual reproduction: kind of reproduction that requires two parents, 65

simple leaf: leaf that has all its leaf blades in one piece, 167

slime mold: protists that have two life stages, 147

species [SPEE-sheez]: group of organisms that look alike and can reproduce among themselves, 131

spontaneous generation [spahn-TAY-nee-us jen-uh-RAY-shun]: idea that living things can come from nonliving things, 37

spore: reproductive cell, 83

spore case: structure that contains spores, 83

stamen [STAY-mun]: male reproductive organ of a flower, 181

stigma [STIG-muh]: top part of a pistil, 193

stimulus: change in the environment, 19

stomata [STOH-muh-tuh]: tiny openings in a leaf, 167

taxonomy [tak-SAHN-uh-mee]: science of classifying living things, 131

temperature: measure of how hot or cold something is, 1

tropism [TROH-pizm]: response of plants to stimuli, 213

tuber: underground stem, 95

vegetative propagation [VEJ-uh-tayt-iv prahp-uh-GAY-shun]: asexual reproduction in plants, 95

vertebrate [VUR-tuh-brit]: animals with a backbone, 233

volume: measure of the amount of space an object takes up, 1

weight: measure of the pull of gravity on an object, 1

zygote [ZY-goht]: fertilized egg, 103